future arquitecturas s.l. 编

万象建筑新闻 5

panorama architecture newspaper

图书在版编目（CIP）数据

万象建筑新闻.5 /《未来建筑》杂志社编. — 天津:
天津大学出版社，2012.11
ISBN 978-7-5618-4549-3

Ⅰ. ①万… Ⅱ. ①未… Ⅲ. ①建筑设计 Ⅳ. ①TU2

中国版本图书馆CIP数据核字（2012）第282880号

责任编辑 朱玉红

出版发行 天津大学出版社
出 版 人 杨欢
地 址 天津市卫津路92号天津大学内（邮编：300072）
电 话 发行部：022—27403647 邮购部：022—27402742
网 址 publish.tju.edu.cn
印 刷 上海瑞时印刷有限公司
经 销 全国各地新华书店
开 本 240mm×320mm
印 张 6
字 数 116千
版 次 2012年11月第1版
印 次 2012年11月第1次
定 价 38.00元

设计师：[uto]厄修拉·弗里克&托马斯·格拉伯纳
导师：帕特里克·舒马赫教授
奥地利因斯布鲁克大学

城市空间/适应性城市建筑

最近刚从奥地利因斯布鲁克大学实验建筑学院毕业的厄修 拉·弗里克(Ursula Frick)和托马斯·格拉伯纳(Thomas Grabner)[uto]在他们的毕业设计中通过数字参数工具对采用倒序合规性设计改造当代城市规划方案的可能性进行了探索。

如果没有设计规划，我们生存的现代社会是无法想象的。社区和城市数量的迅猛增长，使缓慢的适应过程已经无法满足当前的需求。因此，这就向我们提出了一个问题，即通过具有更好适应性的规划理论是否能够推动或模拟从整体上看更加自然的适应过程…… ——弗雷·奥托(Frei Otto)

这项工作通过参数化城市建筑规划对奥托有关城市规划理论的未来发展方向进行了检验。城市理论建议将奥托提出的自然系统和无规划社区生成原则转变为一种紧急都市化开发新形式。无论是生物学的、化学的还是物理学的，在这些自然进化系统内部，一项基础结构原则成为生成智能图案的基础，如同树叶的供应系统、气泡、泡沫或裂缝等图案。检验结果可能并不会被当做典型项目或可预测项目，但为我们提供了一种具有适应性和增长能力的优化解决方案。

该项目位于印度纳威孟买，主要研究与区域和连接有关的数字生成系统与经济和生态之间的关系。通过脚本等数字媒体建立一套基本原则，这样就能实现一种能够解读和反映这些力之间的区域影响的城市肌理，不但能够形成一种固定形式的城市规划方案，而且能够创造出一个具有适应性和显著区别的系统。

designers: [uto] Ursula Frick & Thomas Grabner
tutor: Prof. Patrik Schumacher
University of Innsbruck | Austria

Urban field / Adaptive urban fabric

The diploma project, by Ursula Frick and Thomas Grabner [uto], recently graduated with distinction from the Institute for Experimental Architecture at the University of Innsbruck, Austria, explores the potentials of reshaping contemporary urban planning through bottom-up, rule-based design, enabled by digital parametric tools.

Our modern times are unimaginable without planning. The growth of settlements and cities is so tempestuous that a slow process of adaptation is no longer possible. This, therefore, poses the question, whether by means of more adaptable planning theories, processes can be promoted or simulated which, seen as a whole, are 'more natural'.... —Frei Otto

This work examines the thesis of Otto regarding the future direction of urban planning theory through the lens of parametric urbanism. The urban theory proposed translates the type of generating principles of the natural systems and unplanned settlements invoked by Otto, into a means of developing new forms of emergent urbanism. Within these naturally evolving systems, whether biological, chemical, or physical, a base structural principle becomes the foundation for intelligent pattern generation, as seen in the supply systems of leaves, bubbles, foams or crack patterns. The result may not be read as typical or predictable, but offers an optimized solution capable of adaptation and growth.
Located in Navi, Mumbai, India it focuses on the interrelationship of digitally generated systems regarding territory and connection, with economic and ecological aspects. By establishing a set of foundational principles through digital mediums such as scripting, the result is an urban fabric capable of reading and responding to the influence of relational fields between these forces, creating not a fixed form of urban planning, but rather an adaptive, differentiated system.

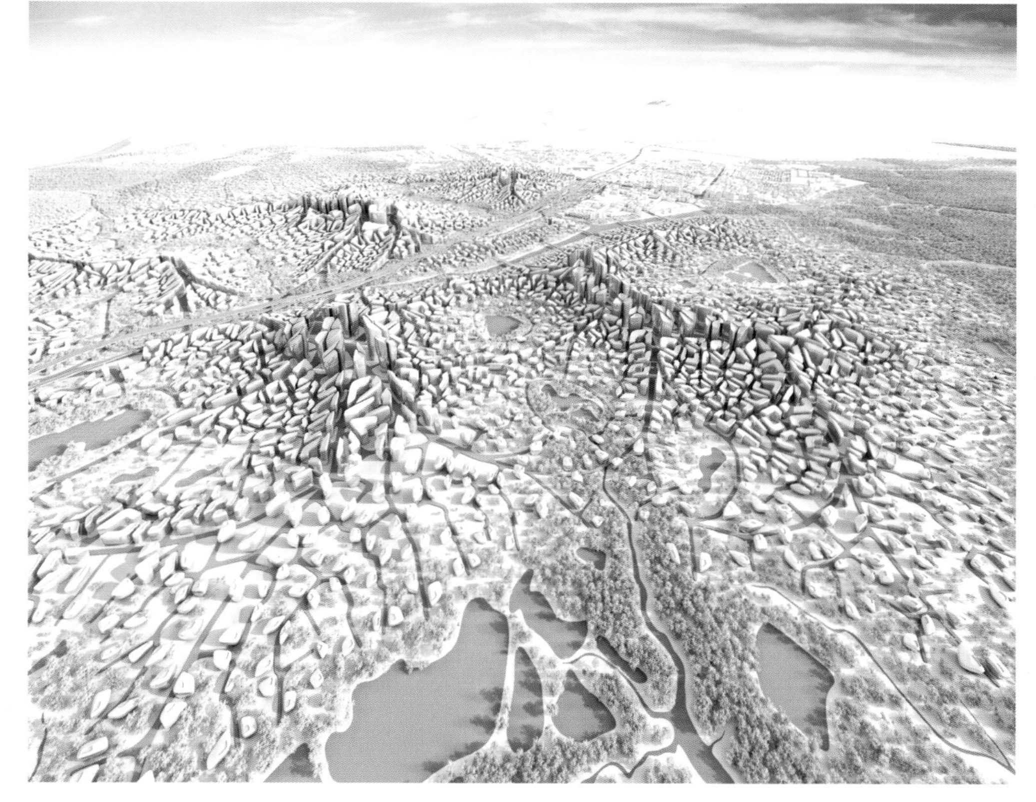

思维景观
THINKSCAPE

与埃利希建筑师事务所负责人的简短对话
A short conversation with EHRLICH ARCHITECTS

Answered by Steven Ehrlich, FAIA, RIBA, Design Principal, Ehrlich Architects

University of California, Irvine Contemporary Arts Center in Irvine, California (pending LEED Platinum)

您是否认为金融危机已经对全球建筑行业进行了重组？

是的，这是绝对的。如今人们更加注重建筑的可持续性和实用性。

您如何看待建筑行业在中国以及其他发展中国家将来的发展前景？

发展中国家有许多建筑项目正在建设，建筑行业在这些国家具有令人兴奋的广阔的发展前景。随着中国以及其他发展中国家的不断发展，我希望这些国家将来能够像拥护现代化一样将他们自己的区域和文化独特性融入到建筑当中。我认为建筑应该既具有全球性又具有区域性。如果先进科技的应用结果使全球的建筑看起来千篇一律，这将是建筑界的一种　悲哀。

拥有华丽外观的建筑时期是否已经结束了？

我希望如此。“嘿，快看我”的建筑时代已经结束了，尤其是当这些建筑的华丽外观以损失功能性和消耗全球资源为代价时。

您是否认为将来建筑行业有某种发展趋势？

建筑正在变得更加环保和实用。我们的预算变得更加紧张，而且正在回归到建筑的实用性和灵活性等方面。建筑必须拥有多种用途以及适应改变的能力。科技可以是孤立的，所以实现一种社会目的对建筑来说非常重要，如将人们集合到一起以及促进人们之间的团结。

您靠什么保持设计热情？

我不断努力，通过打造高品质建筑来满足客户和社区的需求与渴望。

生态学和可持续性在您的作品中起到怎样的作用？

我们尽最大努力建造最具可持续性和最为节能的建筑。其中一种方法就是建造经久耐用的建筑。我们采用在本地建筑中已使用数百年的被动型策略来应对区域气候。

选择：严肃或者文雅？

两者兼具！

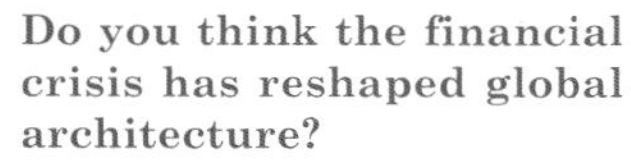

Do you think the financial crisis has reshaped global architecture?

Yes, absolutely. There is a much greater emphasis now on sustainability and practicality.

How do you see the future in booming countries such as China? And in the rest of the world?

The future is tremendously exciting in booming countries where there is so much building going on. As China and other developing countries march forward I hope that they will incorporate their own regional and cultural uniqueness into their architecture, even as they embrace modernism. I think that architecture should be both global and local. It would be a shame if the result of advanced technologies were that buildings came to look the same all over the world.

Are the days of ostentatious architecture at an end?

I hope so. The time for "Wow, look at me" architecture is over, in particular if it's at the expense of functionality and the consumption of global resources.

Do you think there is a trend for the architecture of the future?

Architecture is becoming much more environmental friendly and functional. We are all living with tighter budgets and with that comes a return to practicality and flexibility. Buildings must often serve multiple purposes and have the ability to adapt to changes. And because technology can be isolating, it is important that architecture also serves a social purpose, bringing people together and reinforcing communities.

What keeps your emotion of design?

I am always emotionally engaged in creating buildings that serve the needs and aspirations of my clients and their communities.

How do ecology and sustainability play a role in your work?

We strive to create the most sustainable, energy-efficient buildings that we can. One way we do that is by constructing buildings to last. We also believe in responding to local climates by adopting passive strategies that have been employed in indigenous buildings for centuries.

Choose: Be political or be polite.

Be both!

Pomona College Sontag & Pomona Residence Halls, Claremont, California (LEED Platinum certified)

焦点
ZOOM

起伏的波浪
Undulating movements

柏布拉市拥有丰富的文化传统，每个角落都清晰展示了该市所经历的悠久历史。在这里，我们能够欣赏到各式各样相互交织在一起的建筑结构、建筑肌理与声音。人们经常把此次竞赛中所要求的纪念碑当做一件雕塑作品，然而该设计方案却创造了一个独特的波浪形空间结构。

着眼于整个城市的设计观点

该项目旨在通过创造新型的公共开放空间来促进对项目用地的有效利用，以及进一步完善优美的城市形象和环境景观。广场上种植了150棵树木，象征着“五月五日独立战争”纪念150周年。

The City of Puebla is a place with a vast richness in culture and traditions where every corner is a clear proof of its history made throughout time; we can find a kaleidoscope of buildings, textures and sounds that are all interwoven. While the competition called for an emblematic monument, often regarded as a sculptural object, this proposal creates an undulating spatial fabric.

A viewpoint that looks over the city

The project aims to activate the use of the site by creating new open public spaces and reinforce the beautiful city views and landscape. 150 trees were located throughout the square, representing the years that mark the anniversary of the Battle May 5th.

TEN Arquitectos/Enrique Norten
"Memorial Battle May 5th", Puebla, Mexico
“五月五日独立战争纪念”，柏布拉，墨西哥

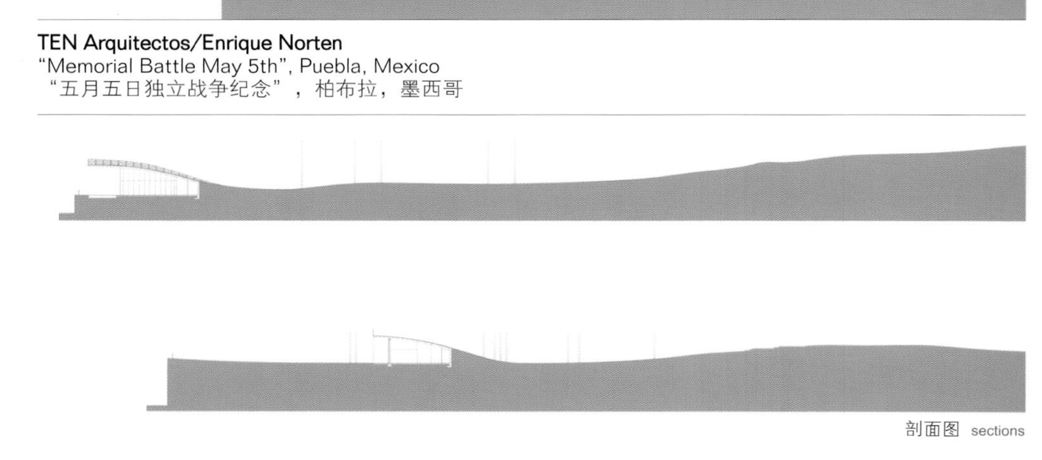

剖面图 sections

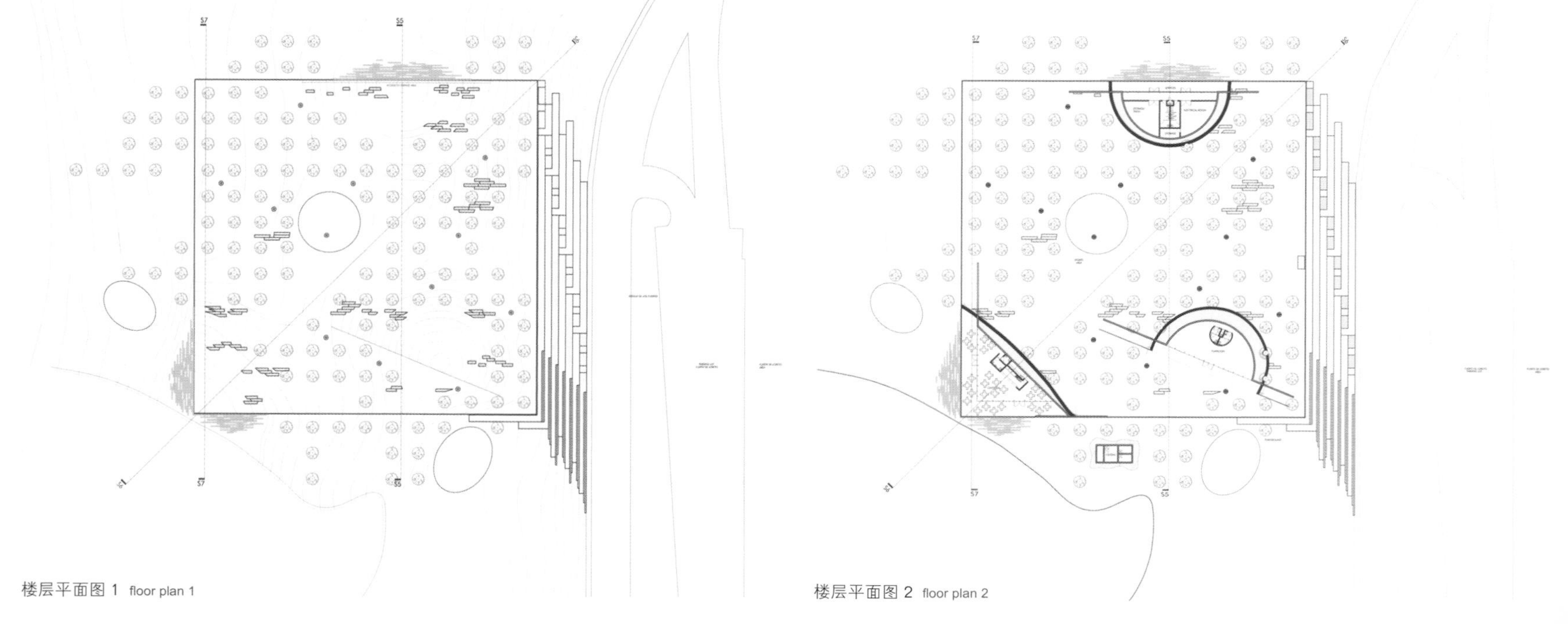

楼层平面图 1 floor plan 1

楼层平面图 2 floor plan 2

生态模型项目

An ecological model project

Laboratory for Visionary Architecture [LAVA]
"Green Climate Fund [GCF]", Bonn, Germany
"绿色气候基金 [GCF]"，波恩，德国

该项目的设计理念来源于莱茵河谷的优美景观。这个三层楼高的结构拥有曲线外观、自然照明天井、屋顶花园以及为餐厅设计的大型沉降式露台，符合最新能源与建筑生态学标准，并且满足可持续性(德国金牌认证)、生态学和能源效率 (净零能耗)等方面的最高要求。

用更少的能耗/工期/成本完成更多建筑

设计师们已经成功地制造出智能系统以及能够抵抗气压、温度、湿度、太阳辐射和污染等外部影响的材料和建筑皮肤。

With a design inspired by the beautiful setting in the Rhine valley, and with curvilinear forms, natural light wells, roof top gardens and a large sunken terrace for the restaurant, the three-level structure will comply with the latest energy and building ecology standards, meeting the highest demands in terms of sustainability (German gold certificate), ecology and energy efficiency (net zero energy).

More (architecture) with less energy/time/cost

They have worked to generate intelligent systems, materials and skins that respond to external influences such as air pressure, temperature, humidity, solar-radiation and pollution.

每个房间都具有独特性
Every room is unique

Heatherwick Studio
"Sheung Wan Hotel", Hong Kong, China
"上环酒店"，香港，中国

这个40层楼高的酒店共有300间客房，所处位置以散装、挂在商店门前和堆放在篮子中的海鲜而闻名。

The forty-storey hotel with three hundred rooms is located in a famous area known for its smell with unpackaged and seafood hanging from shop fronts and piled high in baskets.

建筑的外立面似由数千个"盒子"组成

External façade is composed of thousands of " boxes "

该项目为室内外空间相结合提供了可能性。该设计理念旨在将酒店客房内的床、窗户、小冰柜以及橱柜等常见物品以四个不同尺寸设计成一系列盒子。

It has been an opportunity to conceive the inside and the outside at the same time. The idea was to interpret the familiar objects found in a hotel room—bed, window, mini-bar, cupboard and a place to keep the iron—as a series of boxes, of four different sizes.

© Masaya Yoshimura

具有本国特色的原创方案
Original and vernacular

Takato Tamagami Architectural Design
"Patisserie Uchiyama", Gunma, Japan
"内山糕点店"，群马，日本

该项目将90多年前建造的一家纺织厂翻新改造成为一家糕点店。锯齿形的屋顶在以前的几次翻新中被隐藏了起来，后来得到维护并且重新使用，与周边的建筑风格形成鲜明对比。认识到该建筑的历史背景后，这种稍微倾斜的屋顶形式成为该项目的设计出发点。

A textile factory, which was built more than 90 years ago, has been renovated and transformed into a patisserie. Previously concealed by past renovations, the "saw-shaped" roof was maintained and celebrated, bringing attention to the prevalent architectural element found in the neighborhood. Recognizing it's historical qualities, this gently sloping form became the point of departure for the project.

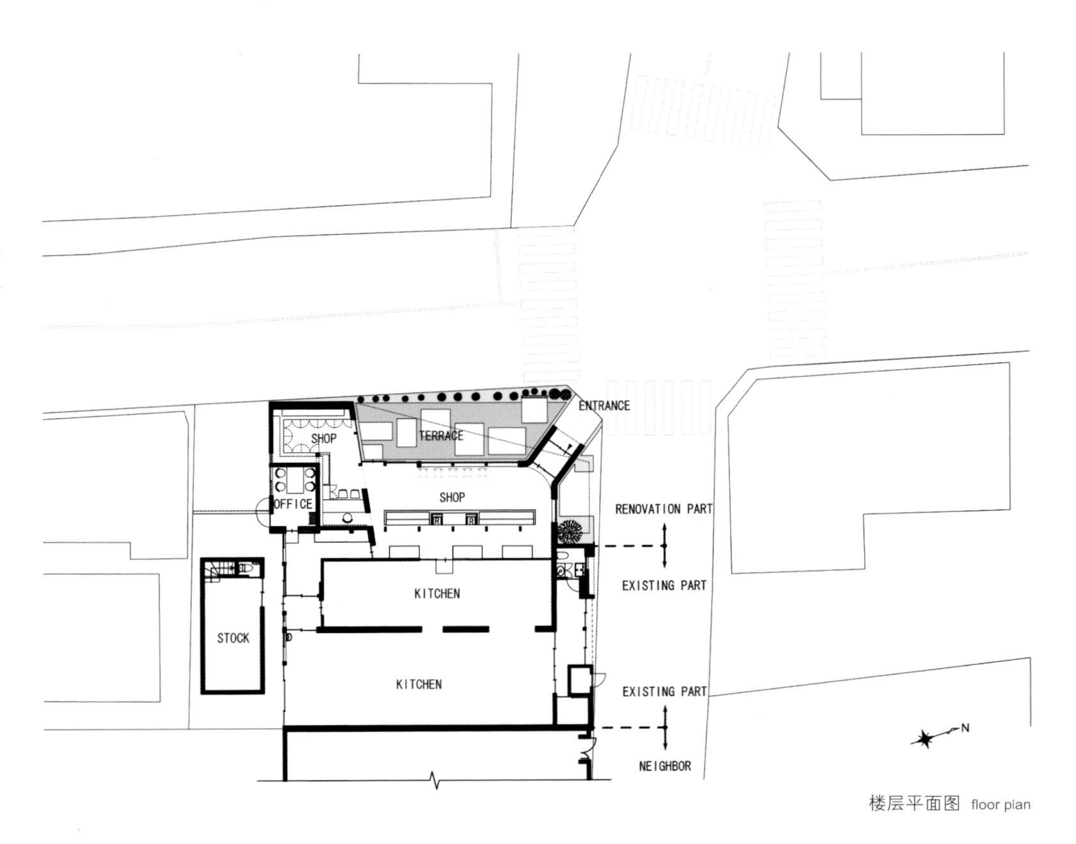

楼层平面图 floor plan

重新结合
Reconnected

OMA
"Rothschild's London Headquarters", London, UK
"洛希尔集团伦敦总部"，伦敦，英国

"新庭"(New Court)是洛希尔集团自1809年以来在伦敦建造的第四个总部项目，每个项目都进一步将圣·史蒂芬教堂(St. Stephen Walbrook)隔离开来。这里起初是城市里两个开放区域，即一个庭院和一个教堂，经过三百年的变化，它们已经融合成为一个毗邻的切个区域。

整个结构向上提升，创造出一条通往高大玻璃大厅的步行通道

在艾伦·凡·卢恩(Ellen van Loon)和雷姆·库哈斯(Rem Koolhaas)的领导下，OMA建筑事务所提出的"新庭"设计方案为整个区域恢复了一种视觉联系。教堂和"新庭"并没有像邻居一样互相竞争，而是形成了一对城市综合体。"新庭"包括一个简单的突出结构，将一系列体块转变成为一种复杂的立方体和附属结构，即一个包含开放办公空间、共享空间和私人工作区的立方体及附属区域。

Rothschild has been located at New Court since 1809. The new New Court is the fourth iteration of Rothschild's London headquarters on the site, each increasingly isolating the church of St. Stephen Walbrook. What began as a dialogue between two open spaces in the city — a courtyard and a churchyard — has, through three centuries of transformation, been reduced to an accidental proximity.

The entire cube is lifted to create generous pedestrian access to the tall glass lobby

OMA's design of New Court, lead by Partners-in-charge Ellen van Loon and Rem Koolhaas, reinstates a visual connection. Instead of competing as accidental neighbours, the church and New Court now form a twinned urban ensemble. New Court is comprised of a simple extrusion transformed through a series of volumetric permutations into a hybrid of cube and annexes: a 'cube' of open office space and appendices of shared spaces and private work areas.

© Matthijs van Roon

待客环境的新概念

A new concept for hospitality

Hofman Dujardin Architects+ Fokkema&Partners
"Eneco headquarters interior design", Rotterdam, The Netherlands
"埃尼科公司总部室内设计"，鹿特丹，荷兰

埃尼科公司总部大楼的室内设计为创造完美的工作环境进行了一次全新的革命。工作区与会议区被设计成充满活力的小岛，漂浮在浅色平静的水磨石地面上。一些小岛采用开放空间的形式，另一些小岛采用封闭式空间供私密性会面，但所有小岛均采用明快的颜色和材料进行装饰。

The interior design for Eneco's headquarters building, has undergone a revolution to create the perfect working environment. The working and meeting areas are designed to be energetic islands floating on an otherwise calm, light-white terrazzo floor. Some islands are open spaces and others enclosed for privacy but they are all executed with vibrant colours and materials.

特殊的身份与环境使人们能够在办公室内更加高效地工作

Specific identities and atmospheres that enable people to orientate themselves better in the office

底层小岛的设计采用红色、紫色和橙色，二楼小岛的设计则采用不同亮度的翠绿色(会议室)和蓝色(工作区)。

Those on the ground floor are red, purple and orange, while those on the first floor are in different shades of verdant green (meeting rooms) and blue (working spaces).

© Matthijs van Roon

© Matthijs van Roon

© Matthijs van Roon

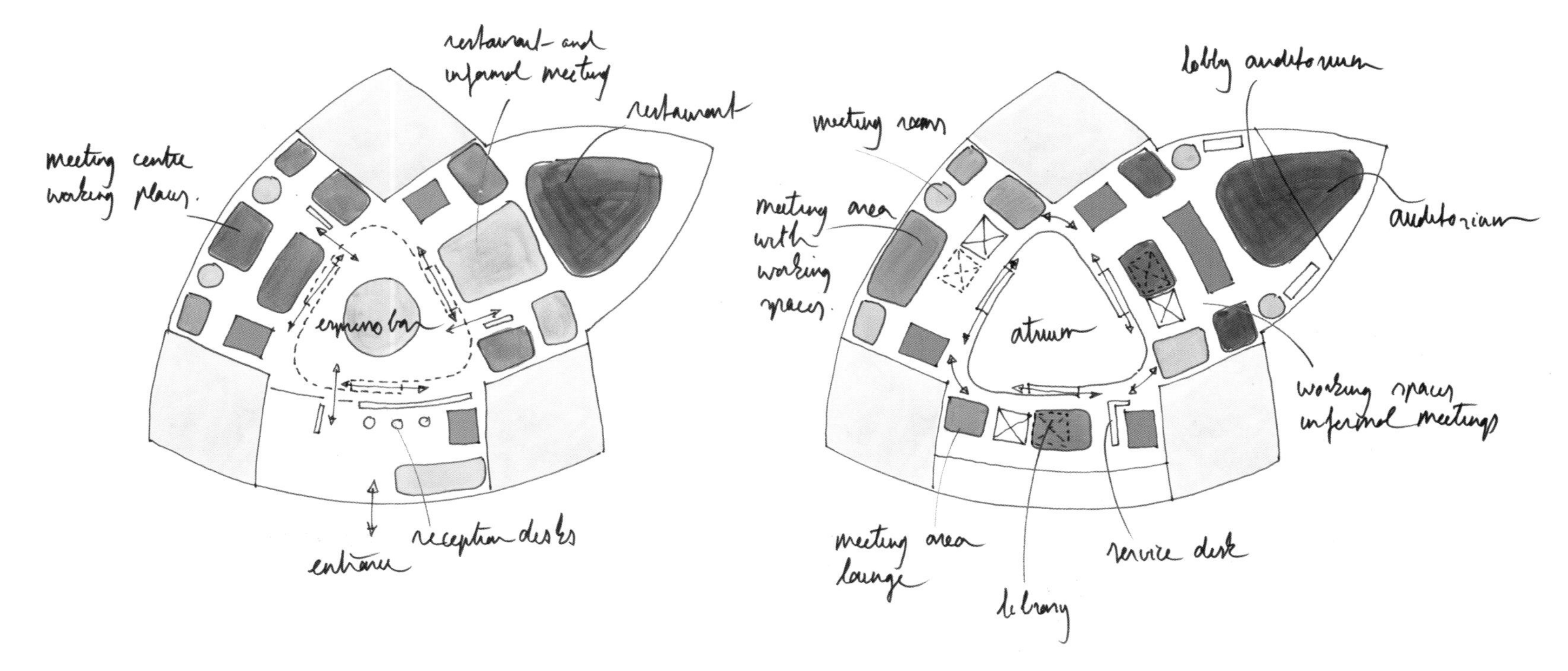

© Matthijs van Roon

三角形几何学
Triangulated geometry

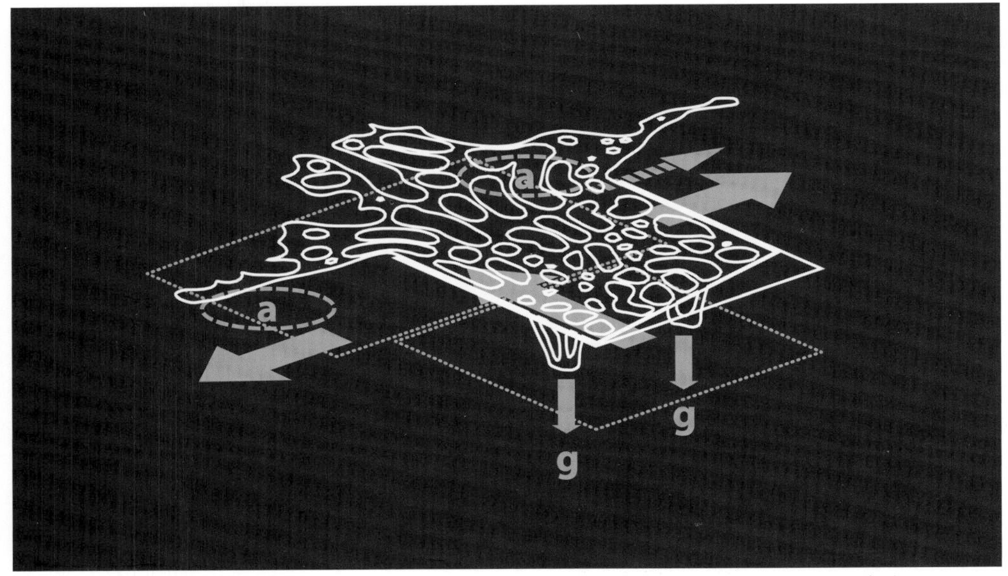

概念图 concept

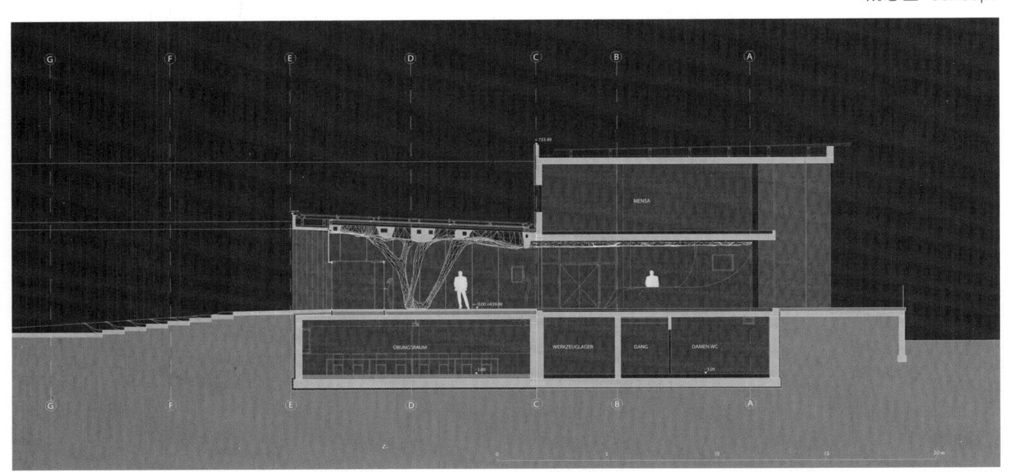
剖面图 section

Soma Architecture
"Foyer Adaption of Building Academy", Salzburg, Austria
"建筑学院门厅改造"，萨尔茨堡，奥地利

该设计方案将一个现有工厂大厅改造成为一个文化建筑。新结构在建筑内外空间之间实现了一种平滑过渡。根据粒子流动原理，利用流体模拟在屋顶上形成一种三维立体样式。流体有三个基本参数，即黏度、密度和表面张力。在计算机上通过一系列变化对上述三种物理性质之间的相互作用进行了测试，形成一种多孔性连贯图案。

门厅将具有多种用途

The scheme adapts an existing factory hall into a cultural venue. The new structure creates a smooth transition between the exterior and the interior of the building. The three-dimensional pattern on the roof was generated in a simulation of fluids based on particle flows. Liquids have three essential parameters: viscosity, density and surface tension. The interactions between the three physical properties have been tested on the computer in a series of variations to generate a coherent pattern with many holes.

The foyer will serve multiple purposes

将自然引入室内

法国东北部低预算可持续性住宅

Bringing nature inside

A low-budget sustainable pavilion in north-eastern France

© Lucas Stoppele

© Lucas Stoppele

St. André-Lang Architectes
"Tourner Autour Du Ried", Muttersholtz, France
"玉米屋"，穆特尔豪兹，法国

该项目为20平方米的圆形住宅，建筑的朝向使室内能够获得最大化的日光照射。建筑结构采用当地材料建造而成。建筑师们采用金属网立面结构，金属网里面储存着大量玉米棒。建筑外观随着玉米棒的轮廓不断变化，此设计灵感来

可持续性与美学特征

自阿尔萨斯平原上使用的玉米烘干机。屋顶设有圆形天花板开口，使明媚的阳光能够进入室内。

The 20 m² circular housing orientated to maximize the sunlight. The structure is constructed with the use of local materials. The architects created a façade of mesh wire as structure, storing countless cobs of corn. The pavilion's appearance will change according to the cycle of corn, taking inspiration from corn dryers in the Alsatian Plains. It features

Sustainable and aesthetically distinctive

a circular ceiling opening allowing entry from the sun's bright rays.

© Lucas Stoppele

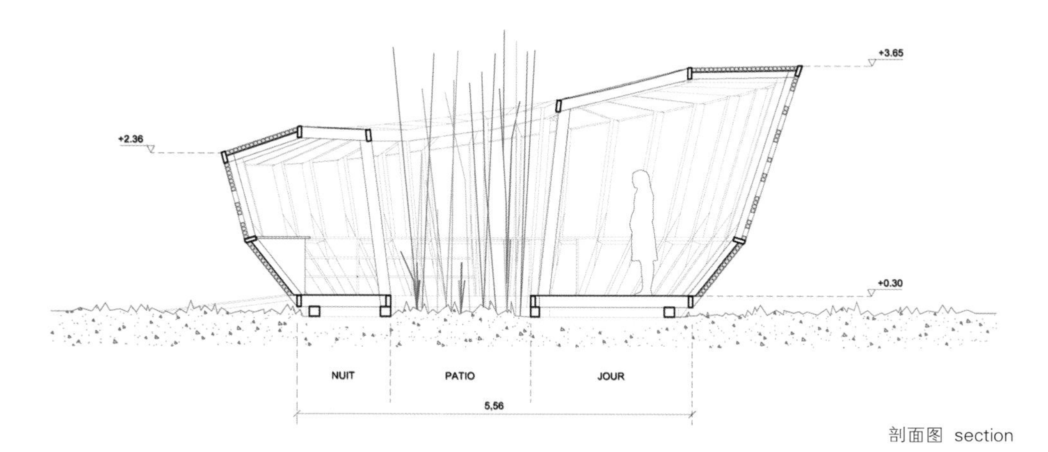

剖面图 section

平面图 plan

© Mecanoo Architecten

垂直木板
Vertical wooden boards

Mecanoo Architecten, Bv
"Maritime and Beachcombers Museum", Texel, The Netherlands
"海洋沙滩博物馆"，太克斯岛，荷兰

太克斯岛(Texel)位于瓦登海(Waddenzee)上，是荷兰瓦登群岛中最大的一个岛屿。每年有大约100万游客光临此岛。

色彩和纹理进一步提高了该建筑的视觉体量

建筑师们为该博物馆设计了四个有趣的相互连接的山形屋顶，屋顶的起伏具有很强的韵律感，从海上望去就像是从堤坝上升起的海浪一样。在数百年的发展历程中，太克斯岛上的居民充分利用搁浅船只或失事船只的浮木来建造房屋和畜棚。Kaap Skil海事博物馆的木质立面就是上述对木材进行回收利用的优良传统的典型实例。

The island of Texel is situated in the Waddenzee and is the largest of the Dutch Wadden Islands. Every year a million or so tourists visit the island.

Color and texture enhance the volume of the building

The museum is designed with four playfully linked gabled roofs which are a play on the rhythm of the surrounding roof tops which, seen from the sea, resemble waves rising out above the dyke. For hundreds of years Texel inhabitants have made grateful use of driftwood from stranded ships or wrecks to build their houses and barns. The wooden façade of Kaap Skil is a good example of this time-hallowed tradition of recycling.

© Mecanoo Architecten

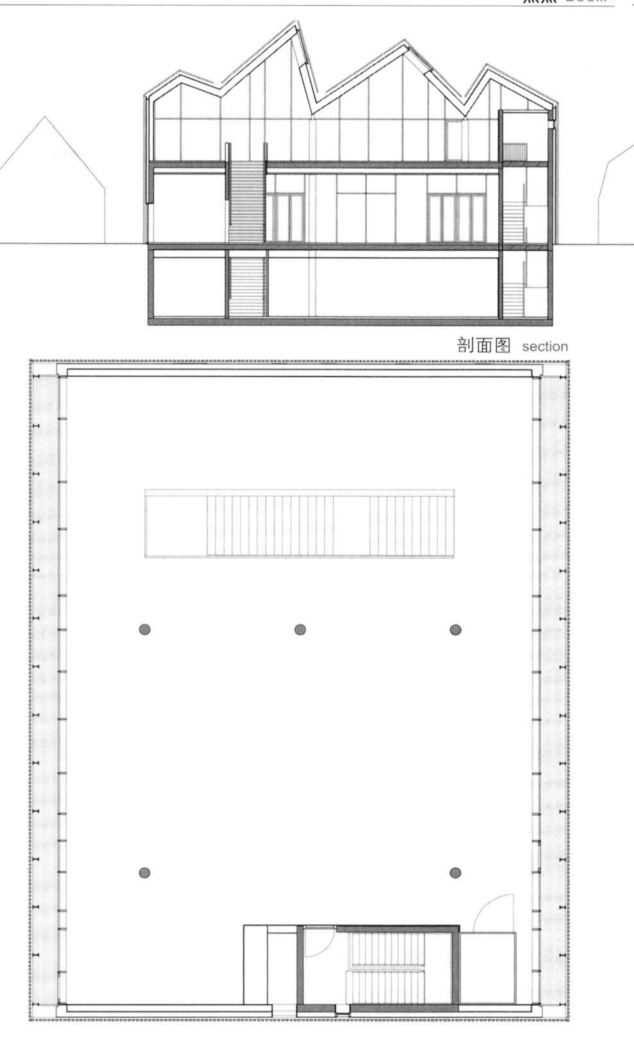

剖面图 section

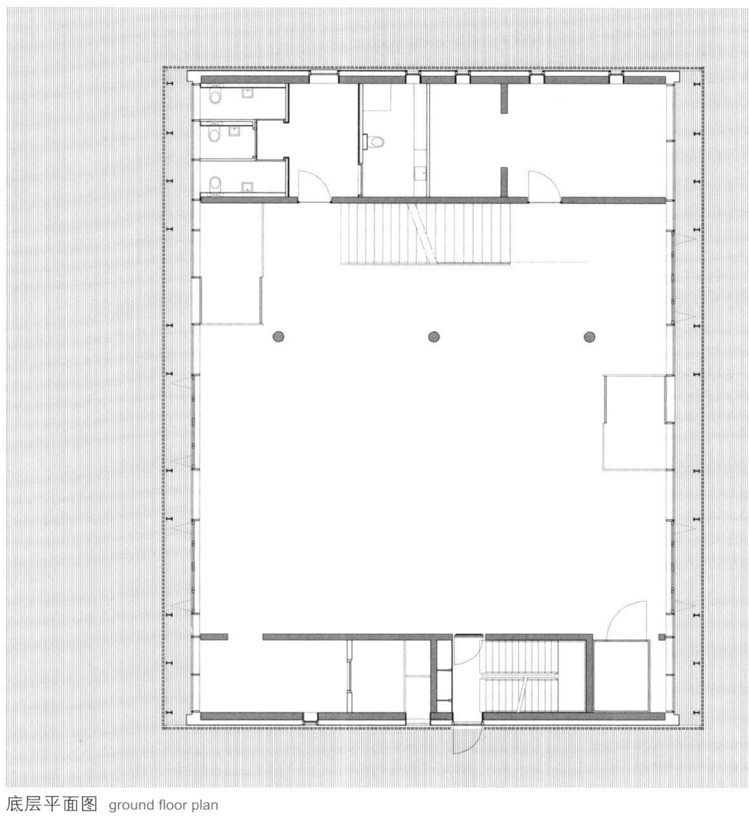

底层平面图 ground floor plan

二层平面图 first floor plan

© Mecanoo Architecten

与景观密切联系

Related to the landscape

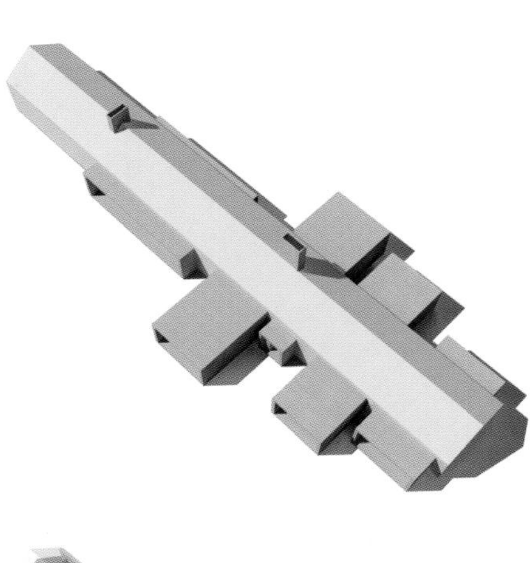

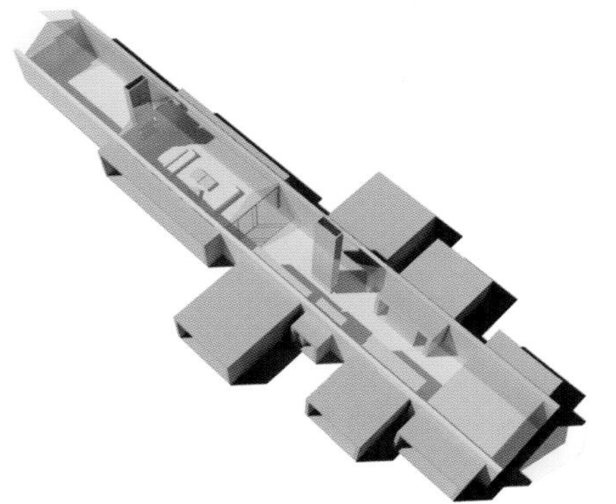

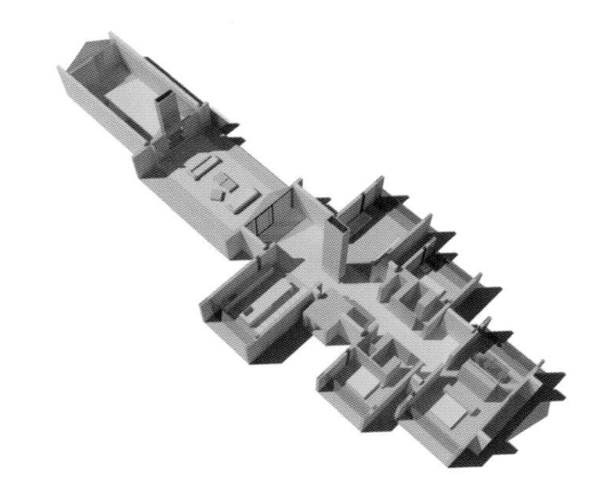

Ramon Esteve
"Refuge", Fontanars, Valencia, Spain
"避难所"，冯塔纳斯，巴伦西亚，西班牙

这个乡村住宅将传统建筑方针应用到新的空间理念当中。斜屋顶的轮廓被挤压成一个线性平面，与周围的小树林形成统一的画面。由松木制成的独立的立方体体块穿插在中心轴的混凝土结构上。

该项目消除了与周围葡萄园和树林之间的界限

该住宅遵循被动式住宅的设计方针。采用太阳能板收集太阳能，将净化雨水作为饮用水，同时采用生物质能源供应系统以及石棉隔热系统，这些技术将大大降低该建筑的能耗，并且提高热质量。

The country house takes the traditional building guidelines to be applied in a new spatial concept. The outline of the pitched roof is extruded into a linear plan to integrate views of the surrounding woods. Separated cubic volumes shaped with pine are attached to the concrete structure of the central axis.

It eliminates the boundary between the surrounding vineyards and woods

The house follows the guideline of a passive house. It will use solar panels to harness renewable energy, rainwater for drinking water, bio-mass energy supply, and rock wool insulation, which will contribute to energy savings and increase of the thermal mass.

© Goettsch Partners

Goettsch Partners
"Northwestern University Music Building", Chicago, USA
"西北大学音乐教学楼"，芝加哥，美国

该项目由美国GP(Goettsch Partners)建筑公司设计完成，将成为西北大学Bienen音乐学院新址，并且在西北大学埃文斯通校园内为交通学院提供更多的教学空间。

该项目加强了对壮丽湖泊的景观视野

新建大楼的面积为1.4万平方米，其中包括一个能够容纳400人的独奏大厅、一个设有150个席位的歌剧排练室/黑匣子剧场以及一个223平方米的合唱团排练室和图书馆。该项目还包括多个教室，教学实验室，教师办公室，供合唱、歌剧、钢琴演奏以及发音练习的教学工作室，练习室，学生休息室以及行政办公室等功能。整个设计强调建筑的可持续性，成功获得美国绿色建筑委员会颁发的LEED银质证书。该项目预计2015年秋季竣工并投入使用。

玻璃和铝材

新教学楼将音乐学院的各种课程集中到一起

Glass and aluminum

The new building will unify the various programs within the School of Music

Architecture firm Goettsch Partners (GP) has designed the signature building that will be the new home of Northwestern University's Bienen School of Music and provide additional space for the School of Communication on Northwestern's Evanston campus.

It enhances the spectacular views of the lake

The new 14,000 m² building features a 400-seat recital hall, a 150-seat opera rehearsal room/black box theater, and a 223 m² choral rehearsal room and library. The project also includes classrooms; teaching labs; academic faculty offices; teaching studios for choral, opera, piano and voice faculty; practice rooms; student lounges; and administrative offices. The building design emphasizes a sustainable approach throughout, with a minimum of achieving LEED Silver Certification from the U.S. Green Building Council.The project is slated to be completed and ready for move-in in fall 2015.

© Goettsch Partners

与周围环境紧密联系的地块
A site for interaction

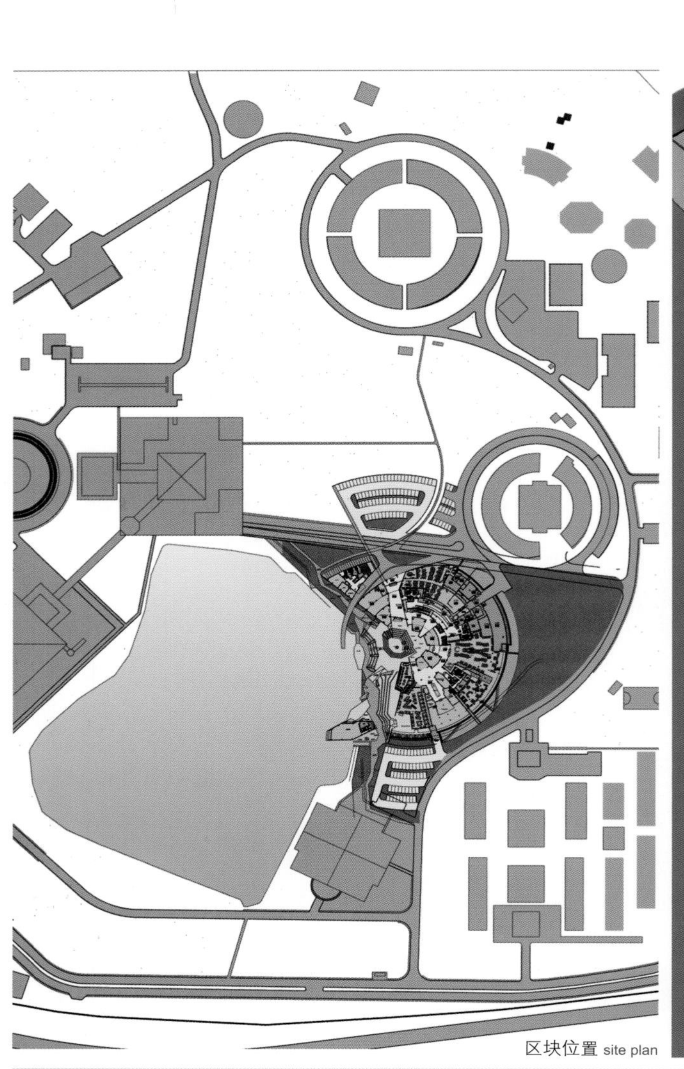
区块位置 site plan

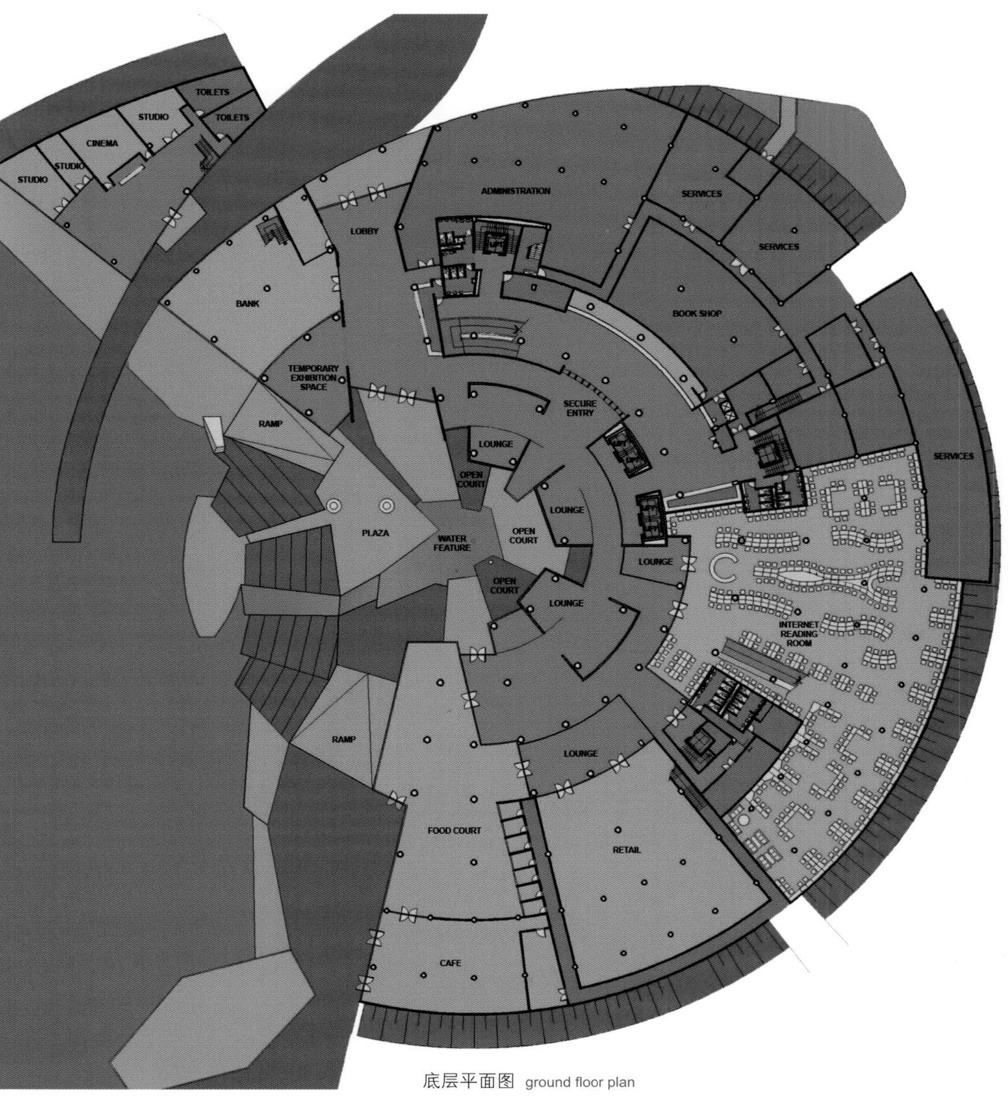

底层平面图 ground floor plan

Denton Corker Marshall
"University of Indonesia Central Library", Jakarta, Indonesia
"印尼大学中央图书馆"，雅加达，印度尼西亚

图书馆位于印尼大学内校园湖泊旁一个非常显著的中心位置，其设计方案通过公开竞争进行评选。该项目的圆形外观与稳重的圆形建筑和道路设计彼此呼应，从校园规划布局中脱颖而出。一座座高塔拔地而起。根据设计理念，这些高塔的设计灵感来自古代印尼人在石碑上镌刻箴言的行为。

Selected in an open design competition, the library is located in a highly visible and central site on the side of the campus lake. Its circular form responds to the strong circular buildings and roadways that distinguish the campus pattern. A series of towers projects from the landform. Conceptually, they take inspiration from the ancient Indonesian practice of inscribing wisdom on stone tablets.

已经成为印尼大学新的学生中心

It becomes the university's new student hub

用花岗岩装饰的不同高度的顶层嵌有狭窄的条状玻璃，使光线能够顺利进入室内空间。

The granite-clad tablets of varying heights are "inscribed" with narrow glazed bands, filtering light into the volumes below.

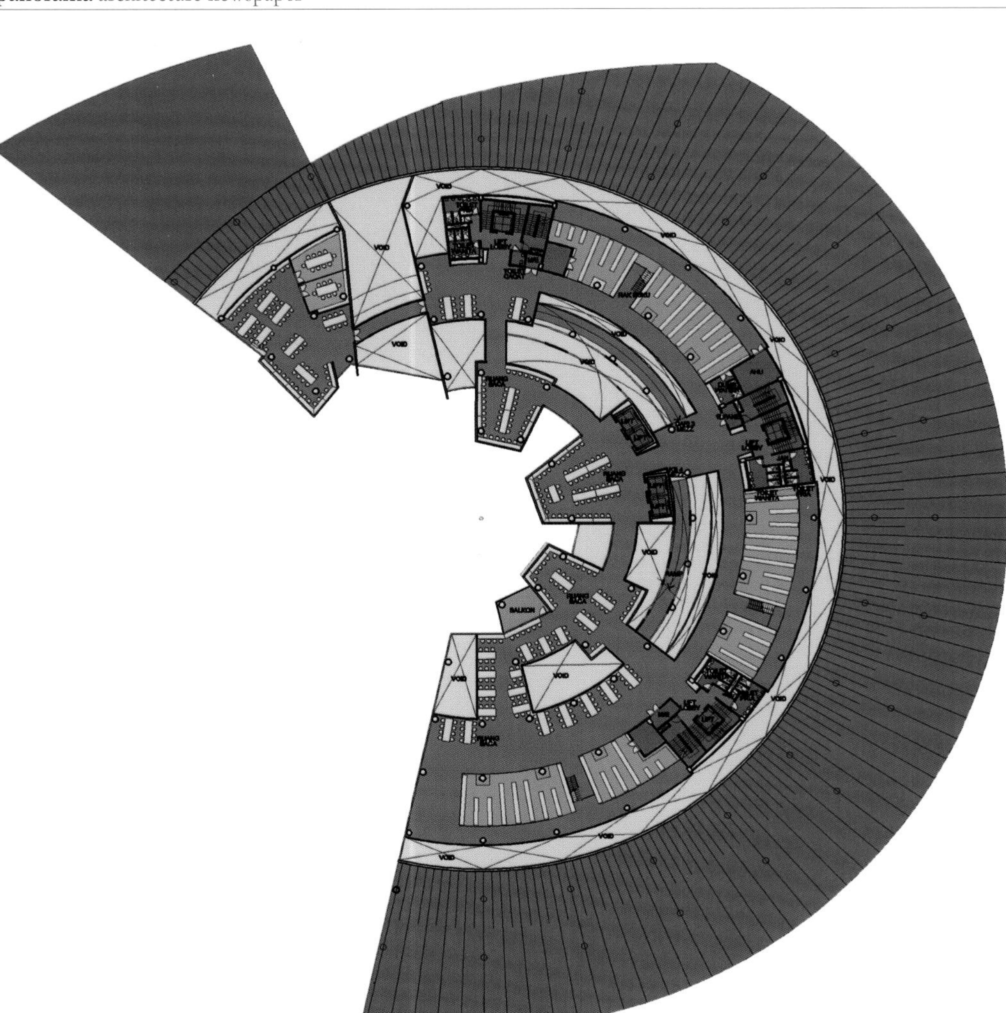

四层平面图 third floor plan

© soma

连续变化馆

世博会结束后该馆将保留下来供游客和当地居民游览欣赏

Continuous transitions

After the EXPO the pavilion will stay as an attraction for tourists and local residents

2012年韩国丽水世博会从2012年5月12日一直持续到8月12日，此次博览会的主题——生机勃勃的海洋与海岸，旨在帮助人们了解海洋与海岸的知识以及先进科技，认识解决海洋挑战的途径和方法。

由奥地利建筑事务所Soma设计的2012韩国丽水世博会主题馆在2009年一次国际公开竞赛中获得第一名。该建筑的空间和结构设计理念以海洋平静而深远的双重特点为基础。连续的多个表面沿垂直方向向水平方向扭曲，形成所有重要的室内空间。

The International Exposition Yeosu Korea 2012 has taken place from May 12 – August 12, 2012. The Expo theme "The living ocean and coast" aimed to help shed light on humankind's knowledge and advancement of technology concerning the ocean and coast and to identify ways to resolve challenges facing the ocean.

The Thematic Pavilion for the EXPO 2012 planned by Austrian architecture office Soma was selected as the first prize winner in an open international competition in 2009. The plain/profound duality of the Ocean has motivated the building's spatial and organizational concept. Continuous surfaces twist from vertical to horizontal and define all significant interior spaces.

© soma

二层平面图 first floor plan

底层平面图 ground floor plan

© soma

吉隆坡最具传统风格的购物中心

Kuala Lumpur's most iconic shopping mall

SPARK
"Starhill Gallery",
Kuala Lumpur, Malaysia
"升禧艺廊"，吉隆坡，
马来西亚

该项目拥有一系列奢侈品店和环境优美的餐厅。该设计方案对升禧艺廊面对武吉免登路又称星光大道(Bukit Bintang)一侧的建筑立面进行了彻底改造。

曲折多变的坚固性与透明度

斯巴克(Spark)建筑师事务所的设计方案将建筑立面打开，在采用玻璃和石板装饰的现有建筑立面上设计连续的商店门面，从而形成大量独特的视觉趣味。

It features an extraordinary array of luxury shops and fine dining restaurants. The design proposal dealt with the reinvention of the existing façade of Starhill Gallery facing Bukit Bintang.

A fractured variation of solidity and transparency

The architects Spark's design has opened up the façade which provides a lot of visual interest via a continuous shop front that wraps the existing building in a crystalline skin of glass and stone panels.

© Lin Ho

© Lin Ho

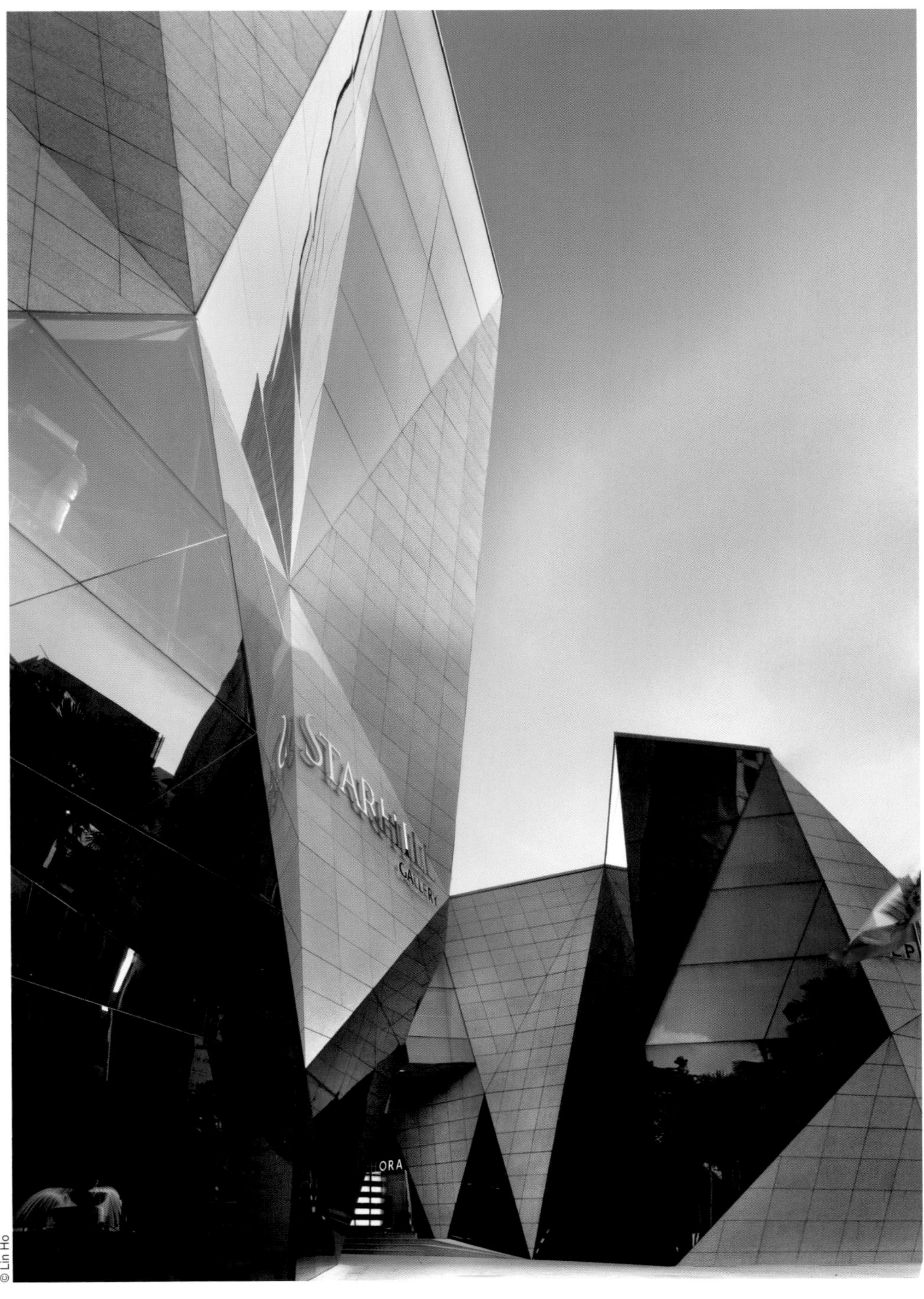

© Lin Ho

Younghan Chung + Studio Archiholic
"Poroscape", Seoul, Korea
"Poroscape时装店"，首尔，韩国

服装地图

A map of clothes

建筑师们在韩国首都首尔零售区仁寺洞路的拐角处设计了一处专门用于服装展示与销售的建筑结构；该建筑外部采用上面布满了小孔的编织砌体单元，人们从设计师们精心设计的开口处能够看到内部的混凝土与玻璃框架。

On the corner of Insadong Road within the retail district of Seoul, Korea, the architects have designed a structure dedicated to clothes; an exterior of woven masonry units features pores to screen and selectively reveals an internal framing of concrete and glass.

这个三层楼高的建筑共191平方米

A total of 191 square meters on three levels

确定建筑表皮等建筑外观元素并不是一个轻松的工作，尤其是在结构改造项目中。该项目的关键点是为现有结构元素(支柱与横梁)重新设计一个新的外观。换句话说，该项目的施工方案以重组为依据(即：拆除现有覆盖物、将整个结构包围起来的同时使原有结构的一部分区域保持外露)。

It is not an easy job to decide on the exterior appearance of architecture, i.e. skin, especially in case of renovation of structures. The main point of this work is to create another look of the existing structural elements (columns and beams). In other words, the architectural method is based on recomposition (tearing down the existing covering, surrounding the structure and exposing a part of the existing structure).

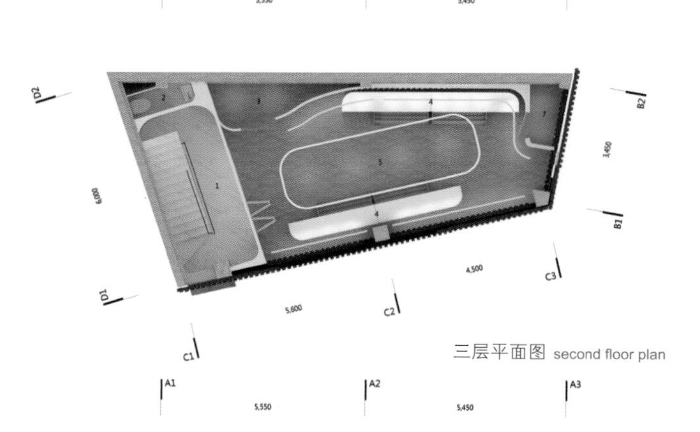
三层平面图 second floor plan

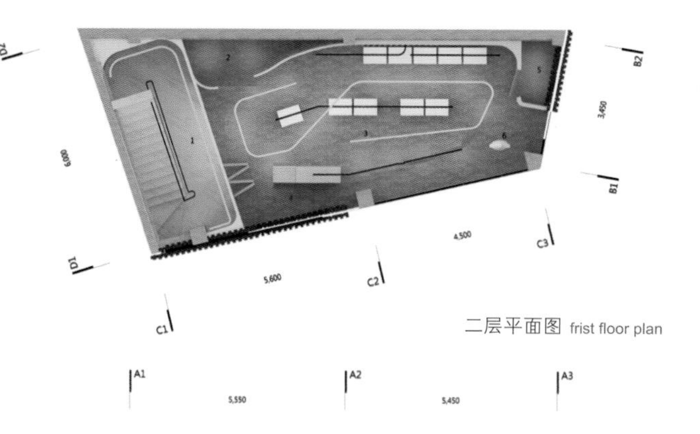
二层平面图 frist floor plan

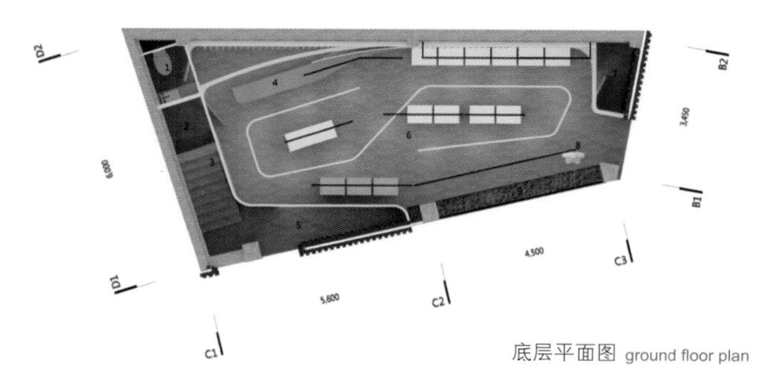
底层平面图 ground floor plan

© Kim Jae Kyeong

© Kim Jae Kyeong

欢快的色调
Vibrant tones

众建筑 People's Architecture Office
"21 Cake Headquarters", Beijing, China
"廿一客办公总部"，北京， 中国

著名美食糕点特许商店的设计巧妙地利用了红、黄、蓝三色的相互作用。这三种颜色的玻璃板互相重叠，在自然光和人工照明条件下营造出一种色彩多变的独特全色谱效果。该项目的会议桌和可移动工作台由众建筑姊妹公司众产品设计事务所制作完成。

The design for a popular gourmet cake franchise relies on the interaction of the three primary colors: red, yellow and blue. The overlapping of glass panels of primary colors creates a full spectrum of changing colors, illuminated by natural and artificial light, with dramatic effects. Conference tables and mobile work tables have been designed and produced by People's Architecture Office's sister company, People's Industrial Design Office.

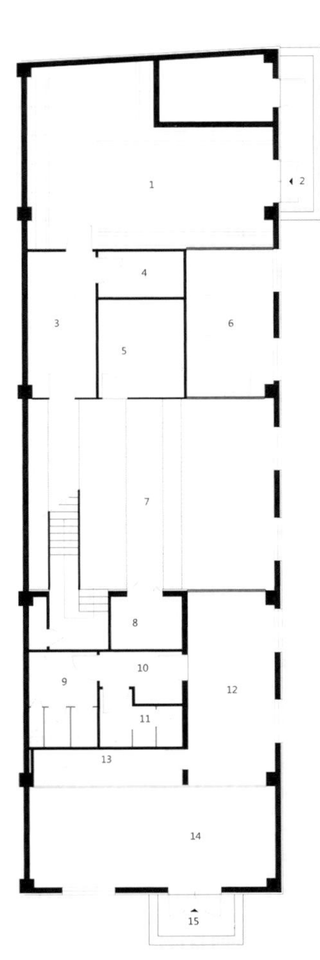

平面图 1 plan 1

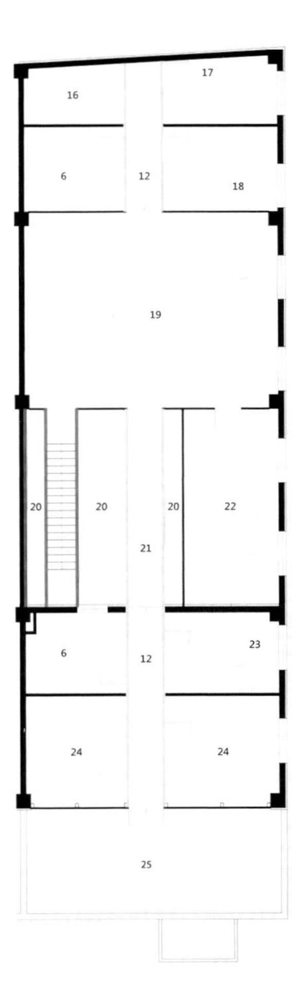

平面图 2 plan 2

电动面板
Electrically powered panels

AS.Archtiecture-Studio
"Novancia Business School", Paris, France
"诺凡西亚高等商学院"，巴黎，法国

法国AS建筑事务所对巴黎始建于1908年的"诺凡西亚高等商学院"进行了扩建改造。在保留原有砖结构、屋檐以及镶嵌图案的基础上，建筑师们利用当代建筑手法为该学院设计了一个崭新的外观。

French firm AS.Architecture-Studio has renovated and extended the original 1908 Novancia Business School in Paris. A contemporary addition gives a new face to the university while keeping the original brickwork, cornices and mosaics.

面板能够明显减少太阳辐射，有效调节室内温度

The panels optimize interior temperature and greatly reduce glare and solar radiation

新的建筑立面由4 102块可旋转矩形薄玻璃板组成，这些玻璃板采用红黄渐变颜色，与旧建筑结构的颜色相呼应。

A new façade made of 4,102 thin rectangular rotating glass shutters with a red to yellow gradient responding to the colors of the old structure.

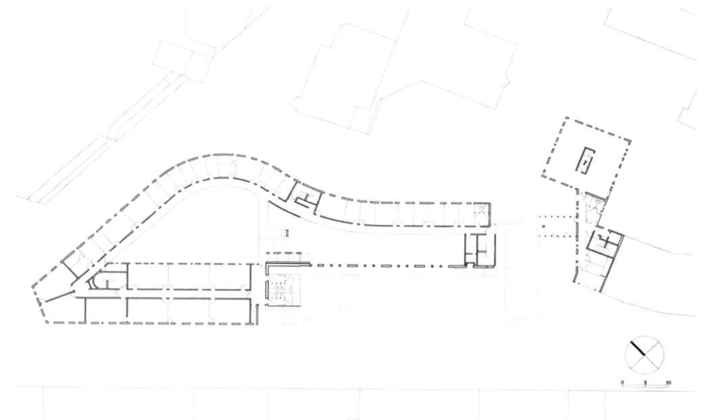

平面图 +3 floor plan +3

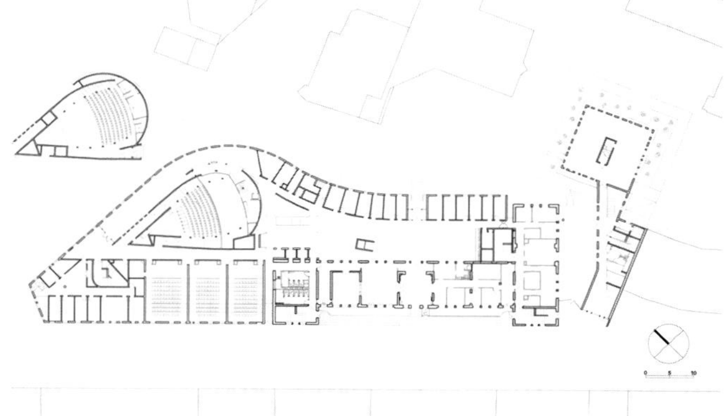

平面图 +1 floor plan +1

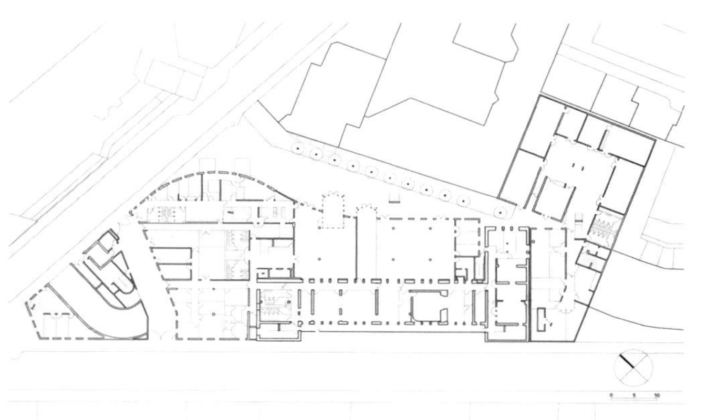

底层平面图 ground floor plan

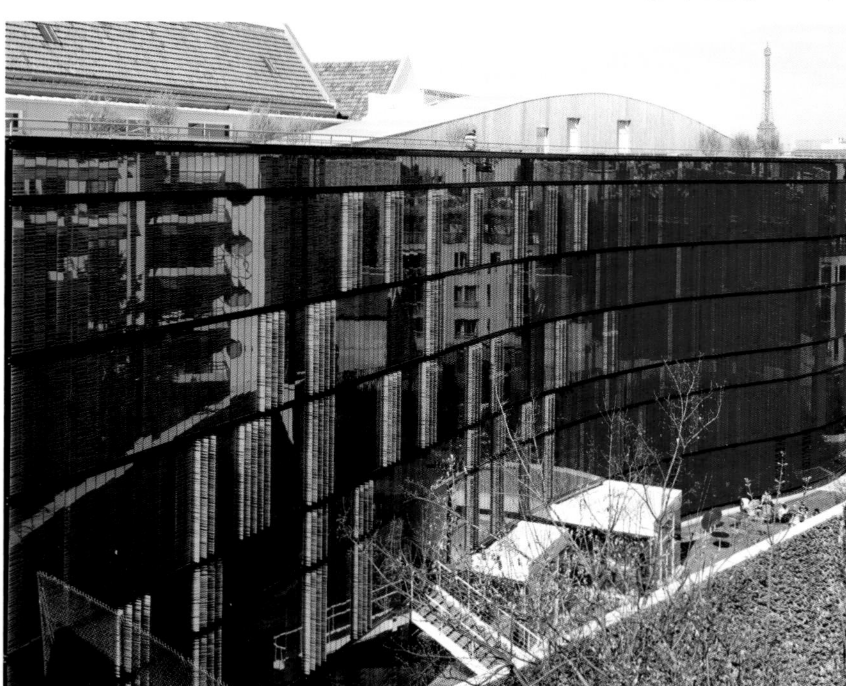

可自由移动

与旅游业相关的新理念

Transportable

A new idea linked with the tourism industry

Studio WG3
"Hypercubus"
"Hypercubus可移动旅店"

该项目专门为不同季节的特殊需求而建，对住宿概念进行了全新的阐释。这个立方体模块适合于奥地利丰富的自然景观，供人们度假时居住所用。该模块充分利用水、电以及景观等既有资源，可以非常方便地搭建、拆除和移动。

旅游者的临时住所

Specifically fabricated to adapt to particular needs depending on the season, the project reinterprets the notion of accommodation. The cubic module was created to provide holiday lodgings that can adapt easily to the fertile Austrian landscape. Taking advantage of pre-existing resources such as water, electricity and landscaping, the modules can be easily installed on site and be up and running in a short amount of time.

A temporary retreat for tourists

立面图 1 elevation 1

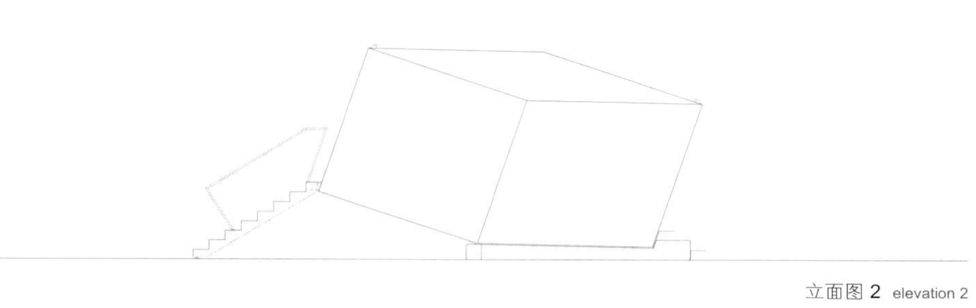

立面图 2 elevation 2

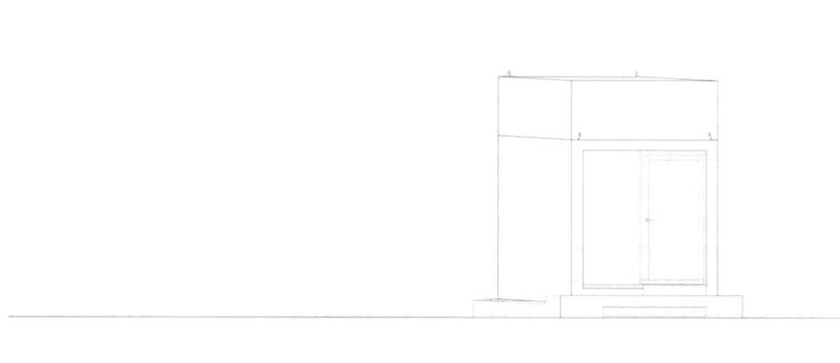

立面图 3 elevation 3

立面图 4 elevation 4

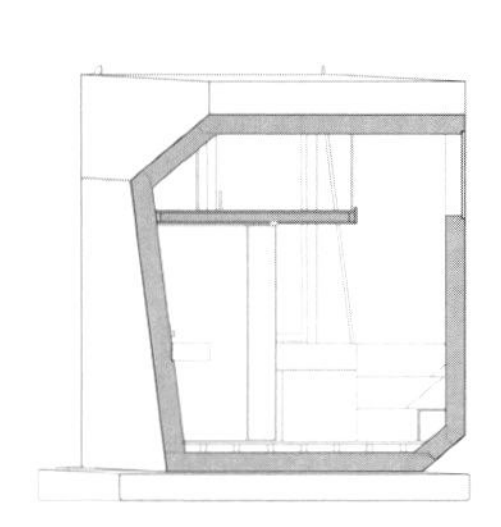

剖面图 AA section AA

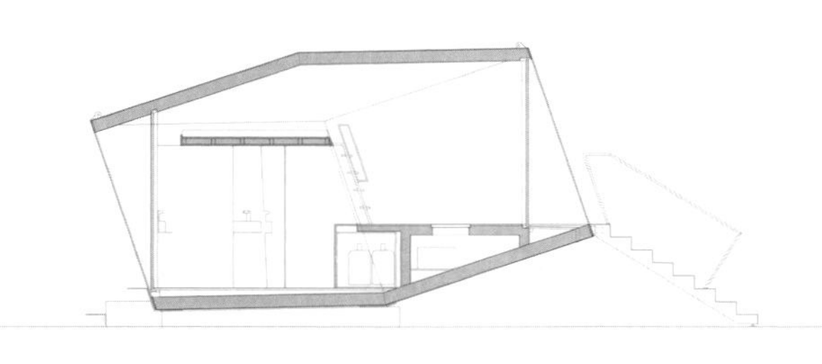

剖面图 BB section BB

具有象征性的历史建筑
A symbolic and historic dimension

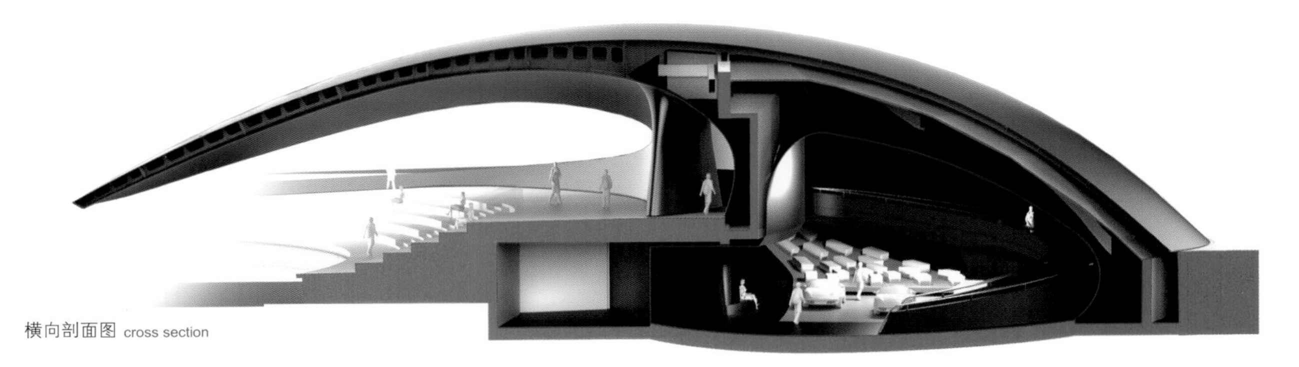
横向剖面图 cross section

© HG Esch

© HG Esch

Henn Architekten
"New Porsche Pavilion", Wolfsburg, Germany
"新保时捷展馆"，沃尔夫斯堡，德国

保时捷厂商的新展馆采用曲线形亚光屋顶设计，该设计理念来自保时捷911车型的外观。

The sports car manufacturer's new exhibition building adopts a curved and gleaming matt roof design, inspired by the silhouette of a Porsche 911.

一个享受感官体验的场所

A place of sensuous experience

这个单壳式建筑结构模拟保时捷长期保持的轻质设计传统来表现"活性表面结构"的设计理念，即：不锈钢屋顶结构采用自承重壳体结构形成内部空间，将展馆的静态元素当做背景元素。

Emulating Porsche's long tradition of lightweight design, the monocoque construction (French: "single shell") employed the "surface-active structure" principle: the stainless steel roof structure is a self-supporting shell shaping the space within, the statics of the pavilion being relegated to the background.

© HG Esch

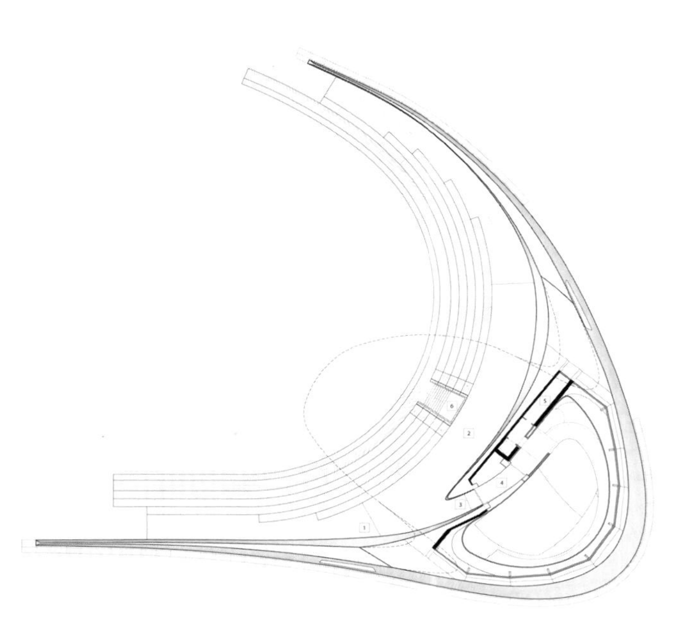

入口层平面图 entrance level plan

© HG Esch

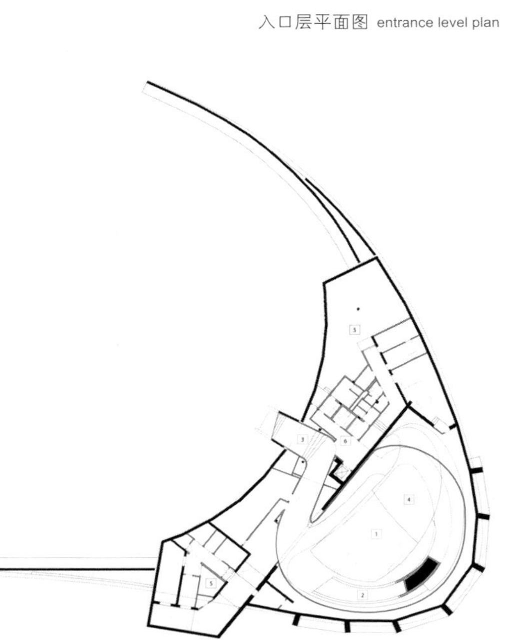

展览层平面图 exhibition level plan

© HG Esch

小型展馆的集合
A family of smaller pavilions

Serie Architects
"London 2012 BMW Group Pavilion", London, UK
"2012伦敦奥运会宝马集团展馆"，伦敦，英国

该展馆位于奥林匹克公园中的水工河(Waterworks River)之上，要求具有某种特定的外观和审美特点。上述要求通过将古典基座重新设想成某些无形或虚幻的结构得以实现。传统的方形基座不但占用空间大，而且非常沉重。而该展馆的基座不但质量轻，而且具有梦幻般的动态效果：水幕从底层屋顶的四周淌下，形成一种不断变化的动态立面效果。

轻质临时性结构

原设计方案将这些展馆在奥运会期间集中到一起，并在奥运会结束之后将它们分散到其他区域。这样，每个展馆都能够在自然环境中找到一个新家。

Positioned directly on the Waterworks River in the Olympic Park the pavilion required a certain presence and aesthetic interest. This is achieved by re-imagining the classical podium as something completely immaterial or ethereal. The traditional plinth is massive and heavy. The pavilion plinth is immaterial, light, and animated: water streams down around the ground floor creating a constantly changing façade.

It is designed to be light and ephemeral

The original architectural conceit was of a group of pavilions huddled close together during the Olympics , but at the end of the Games dispersed to other locations. Each pavilion would find a new home within a natural setting.

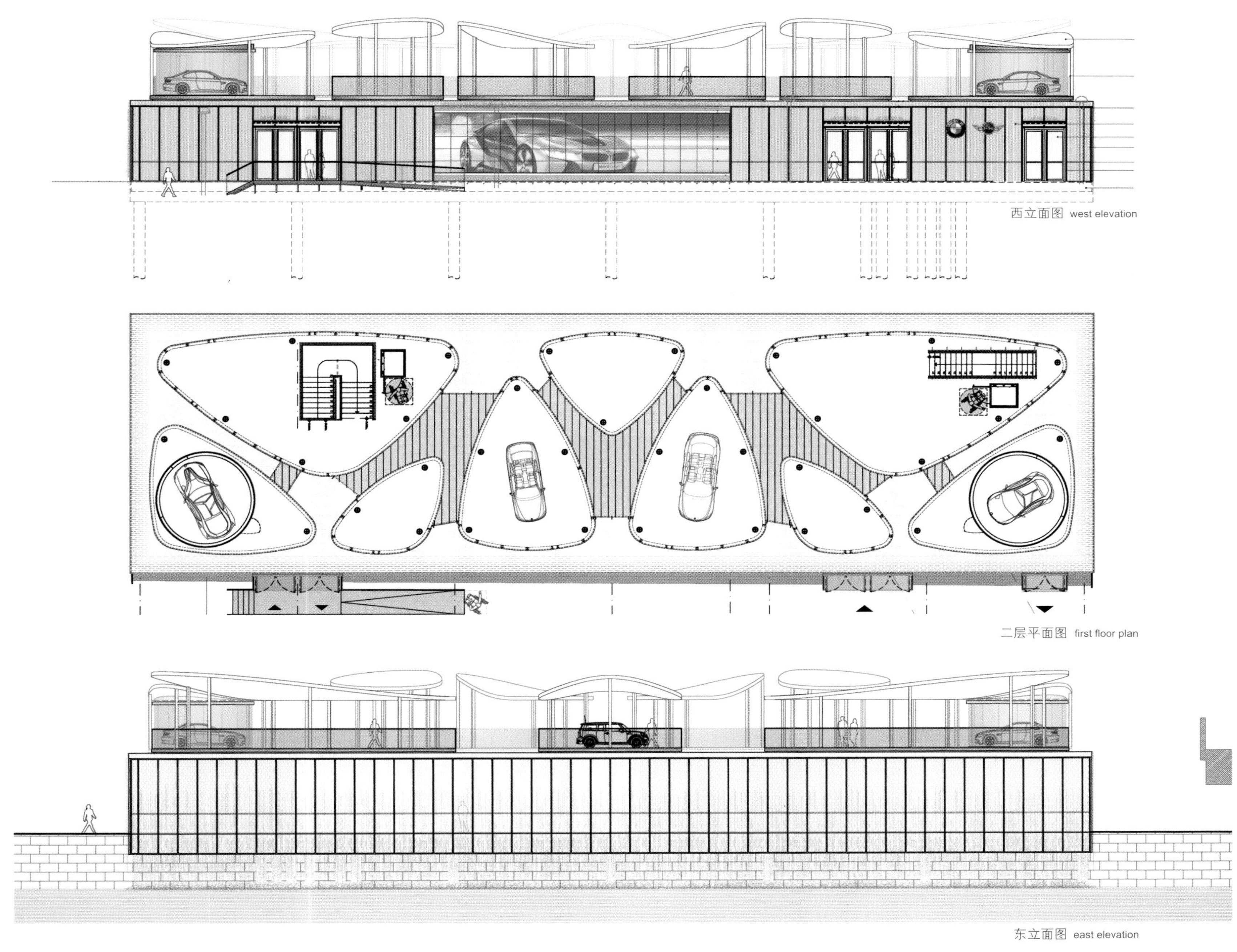

西立面图 west elevation

二层平面图 first floor plan

东立面图 east elevation

纪念碑式的建筑
The antithesis of monumentality

Dr Margot Krasojevic
"Observatory Art Museum", Buenos Aires, Argentina
"天文艺术博物馆"，布宜诺斯艾利斯，阿根廷

项目地块毗邻工业仓库、装卸码头、商业区和自然保护区。该博物馆的设计努力将建筑视图融入到城市空间中，强调真实事物的定义正在不断扩大的现实，从而使观看者、展览空间、城市建筑以及周围环境之间的界限变得模糊。

看上去该建筑马上将要起飞

建筑形式没有空间层次的划分，成功创造出一种漂浮在空中的视觉效果。建筑结构由一个预制激光切割的铝质壳体组成。

The area is a juxtaposition of industrial warehouses, shipping docks, commercial district and nearby nature reserves. The museum's design attempts to arrange images and views into the city to highlight the ever expanding definition of what is considered real, diluting the edges between the viewer, exhibits, city fabric and it's immediate context.

It seems to float over the city

The form has no spatial hierarchy creating an ethereal presence. The structure consists of a single, prefabricated, laser cut aluminium, semi-monocoque shell.

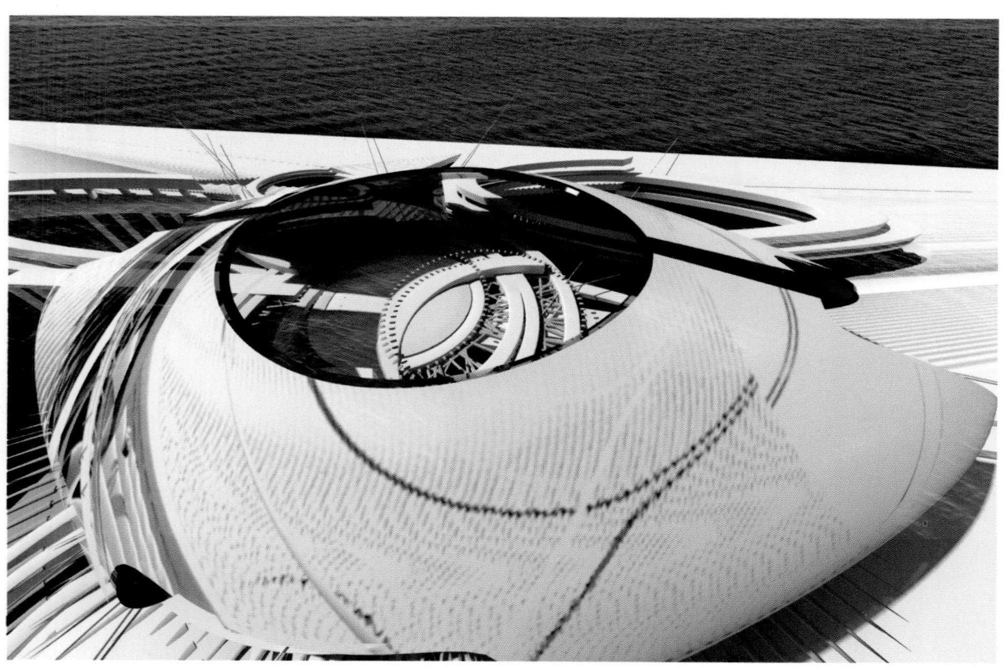

全景视窗
Panoramic windows

3XN Architects
"New Cultural Center",
Mandal, Norway
"新文化中心"，曼达尔，挪威

在尊重城镇历史和周围景观环境的基础上，建筑师们成功创造了这一现代化外观造型。拱形外观造型一部分参考曼达尔周围的小土丘，另一部分参考以前位于该地块的工业中心。

绿化屋顶进一步与周围自然景观融合到一起

该建筑的白色外观与周围原有白色木质住宅相呼应，这些白色木质住宅是曼达尔地区著名的建筑特色。除了该文化中心建筑的设计，3XN建筑事务所还为该地区一座拱桥的总体规划项目提供了设计图，这座拱桥将与该文化中心相连，横跨河流之上。

The modern expression is created with a deep respect for the history of the town and the surrounding landscape. The arched-shape refers partly to the soft hills, located around Mandal, and partly to the industrial center, which previously was located on the site.

The green roof will increase integration with the surrounding nature

The white appearance corresponds with the old white wooden houses, which Mandal is known for and which gives the town its own character. Besides the design of the architecture for the cultural center 3XN has delivered the graphic design for an arch, the area's master plan and a bridge that will go from the cultural center and over the river.

循环交通新体验

A new circulation experience

LYCS Architecture
"Jiaxing University Library & Media Center", Jiaxing, China
“嘉兴学院图书馆与多媒体中心”，嘉兴，中国

这个4.2万平方米的图书馆建筑周围拥有茂密的树木和优美的湖景，并且拥有不同的层次划分，如多个私人空间层次、空间体验层次、内向和外向层次等。

Surrounded by rich woods and luscious water, the 42,000m² library plays with the hierarchy of multiple private spaces; the hierarchy of the pace of spatial experience; and the hierarchy of introversion and extroversion.

建筑中心周围设有一道缓坡，为人们提供了一条更加舒缓的循环通道

A gently sloping ramp wraps the core to create a softer circulation

通过铝材/玻璃立面对光线的反射以及平滑的湖边曲线成功实现了该建筑漂浮晶体般的外形设计。

A floating crystalline form is achieved by the juxtaposition of the glossy reflection of the aluminum/glass façade and the gentle curves of the lakeside.

© LYCS Architecture

© LYCS Architecture

© LYCS Architecture

UNStudio
"V on Shenton", Singapore
"珊顿V形综合体"，新加坡

玻璃六边形建筑
Glass hexagons

为房地产开发商UIC设计的"珊顿V形综合体"(V on Shenton)将代替现有的一座塔楼建筑，该公司自20世纪70年代就在该塔楼中办公。

该项目包括办公和住宅两座塔楼，位于该区域内一个独特的位置，建筑的体量设计也充分显示了其地理位置的优势。办公楼的设计与住宅楼对面的区域相呼应，而住宅楼采用高层设计从周围建筑中脱颖而出。办公楼和住宅楼的立面设计采用同类图案。

最新能源高效型设计理念

设计师采用六边形的基本形状在建筑立面上创造出优美的图案，进一步提升了立面装饰效果。此外，立面上设置的棱角和遮阳装置也适用于新加坡的气候条件。

The V on Shenton building for property developers UIC will replace an existing tower block that has housed the company since the 1970s.
The dual programming of office and residence is a unique situation in this area and the massing of the towers is designed to reflect this. The office tower corresponds to the scale of the area opposite the residence tower, which rises up to distinguish itself from the surrounding buildings. The office and residential façades originate from the same family of patterns.

It features the latest energy efficient design

The basic shape of the hexagon is used to create patterns which increase the performance of the façades, with angles and shading devices that are responsive to the climatic conditions of Singapore.

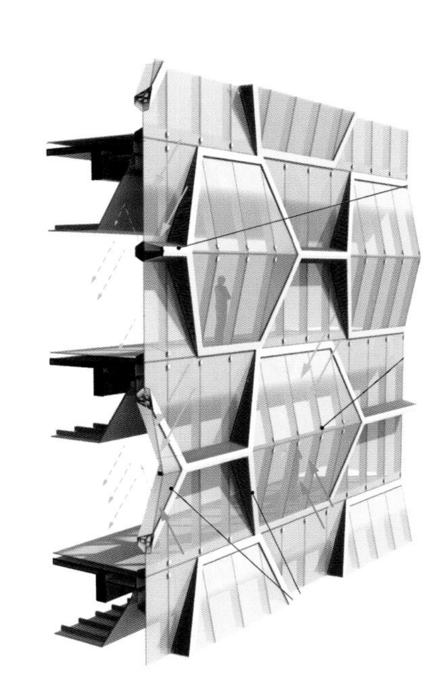

交互式建筑特点
Interactive nature

拾稼设计 IO DESIGN
"Summer International Retail & Entertainment Centre", Zhuhai, China
"夏季国际零售与娱乐中心"，珠海，中国

该多功能开发项目包括36万平方米的租赁零售空间以及办公空间、酒店公寓和住宅单元等。

The mixed-use development contains 360,000 m² leasable retail space together with commercial, hotel serviced apartment and residential accommodation.

丰富而多样化的环境

Rich and diverse environment

该项目地块位于珠海这一快速发展城市的重要区域，此处是城市空间与周围山丘等自然地理环境的交会处。

The site of the development is unique in this important growing city, as it is the meeting point between the grid of the city and the natural topography of the surrounding hill range.

模块化六角形建筑

Modular hexagonal structures

Osamu Morishita Architect & Associates
"Aron R & D Center", Aichi, Japan
"艾隆研发中心"，爱知，日本

该项目以六边形图案为基本设计元素，通过一系列能够随意重组的灵活空间实现便捷的通道系统。

一系列具有微气候调节功能的体块

屋顶设有出风口，帮助调节室内温度；外层防水膜能够收集雨水进行循环利用；其间点缀的绿化庭院和高塔能够降低周围的气温。

The project adopts hexagon patterns as foundamental design element and fosters free-flowing circulation through a series of flexible spaces that allow for constant re-adaptation.

A set of volumes climatically controlled

A designed air outlet on the roof helps control internal temperatures while the outer waterproof membrane collects rainwater for recycling. The interspersed green courtyards and towers help to reduce ambient air temperature.

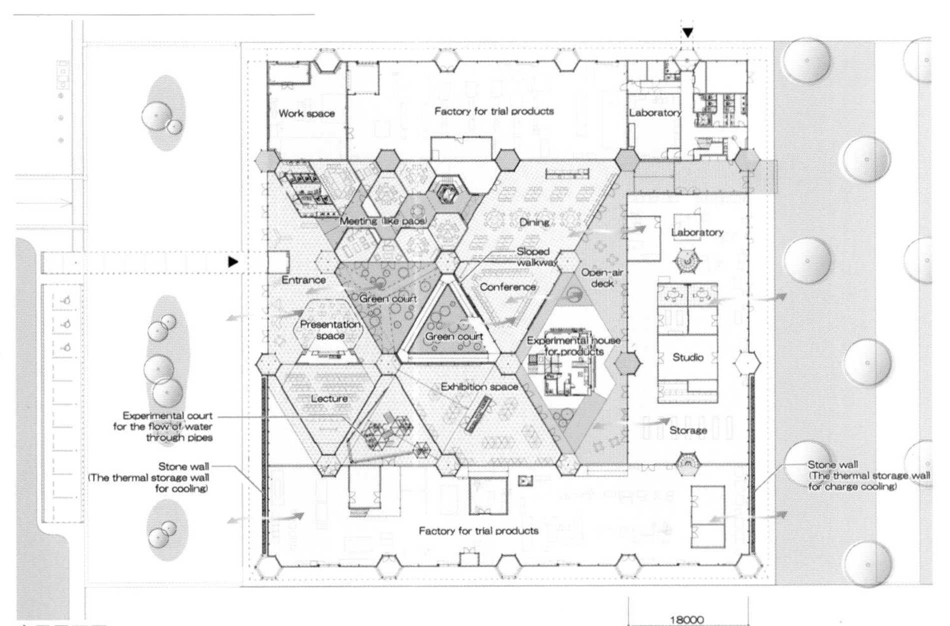

底层平面图 ground floor plan

二层平面图 first floor plan

椭圆形结构

An oval-footprint structure

NL Architects (Pieter Bannenberg, Walter van Dijk, Kamiel Klaasse)
"Bicycle Club", Sanya, China
"自行车俱乐部"，三亚，中国

万科地产委托阿姆斯特丹NL建筑师事务所设计一个自行车俱乐部，作为公司目前在华南地区一个大型项目的一部分，要求该俱乐部具有自行车租赁和咖啡馆等服务功能。

Housing Corporation VANKE has asked Amsterdam-based NL Architects to make a proposal for a Bicycle Club as part of a big resort in Southern China that they are currently involved in. The project should accommodate bicycle rental and a cafe.

外延式屋顶是热带气候对建筑的基本要求

A protruding roof is essential in this tropical climate

超大型屋顶设计能够提供额外的功能。在这里设计一个室内自行车赛车场怎么样？椭圆形轨道拥有陡峭边缘和高雅曲线，形成一种乐观的姿态；向上卷曲的屋檐使人们联想到具有强大功能的宝塔。

The oversized roof perhaps could house an additional function. What about a velodrome? The elegant curvature of the steeply banked oval bike track creates an optimistic gesture; eaves curled upward evoking a surprisingly functional pagoda.

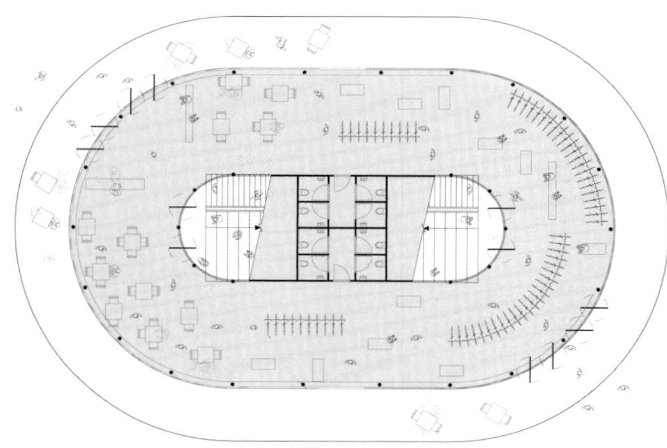

平面图 plan

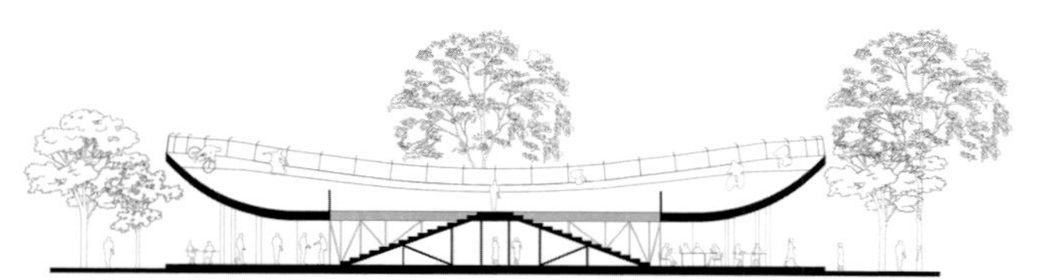

剖面图 section

完美的共生系统
A perfect symbiosis

Ross Lovegrove
"Solar Tree",
"太阳能树灯"

这款"太阳能树灯"由阿特米德(Artemide)公司设计，设计灵感来自自然界中的有机形态，对树木的形态进行了全新的阐释，并且将大自然的敏感性引入到城市环境当中。

它代表着我们时代的基因

这是一棵体态高雅并长有生态智能"果实"的高大树木，太阳能板白天收集太阳能并在晚上向LED灯供电。

Solar Tree by Artemide, draws inspiration from the organic forms of nature, reinterpreting the morphology of the tree and introducing the sensitivity of the natural world into the urban context.

It represents the DNA of our time

A elegant tree with ecologically intelligent "fruits" – that is, the LED bubbles that light up at night powered by the solar energy accumulated during the day by solar panels.

钢丝带
Steel ribbons

Molo Design
"Nebuta House",
Aomori, Japan
"睡魔节文化中心"，青森，日本

该建筑是一个有关睡魔节(Nebuta)各方面文化以及日本北方城市青森县创造性文化的博物馆和文化中心。睡魔祭(Nebuta Matsuri)是日本三个最著名和最大规模的节日之一，是一种讲故事模式的祭祀节日。在节日期间，人们将历史和神话故事中的英雄、恶魔和其他生物做成大型(9m×7m×5.5m)的纸灯笼(即睡魔)。该建筑外墙包裹着扭曲的钢带，并且设置了开口用于不透明区域、景观或步行通道的采光。

钢带幕墙立面成功创造了一个叫做“缘侧”的遮蔽式露天全周空间

The building is a museum and culture centre dedicated to all aspects of the Nebuta festival and its creative culture in Aomori,the Northern Japanese city . Nebuta Matsuri is one of the three most famous and largest festivals in Japan with form of storytelling .During this period heroes, demons and creatures from history and myth come to life as large-scale (9m × 7m × 5.5m) paper lanterns (Nebuta). The building is enclosed by ribbons of twisted steel, with openings for light, areas of opacity, views, or opportunities for pedestrian circulation.

The ribbon screen façade creates a sheltered outdoor perimeter space called the "engawa"

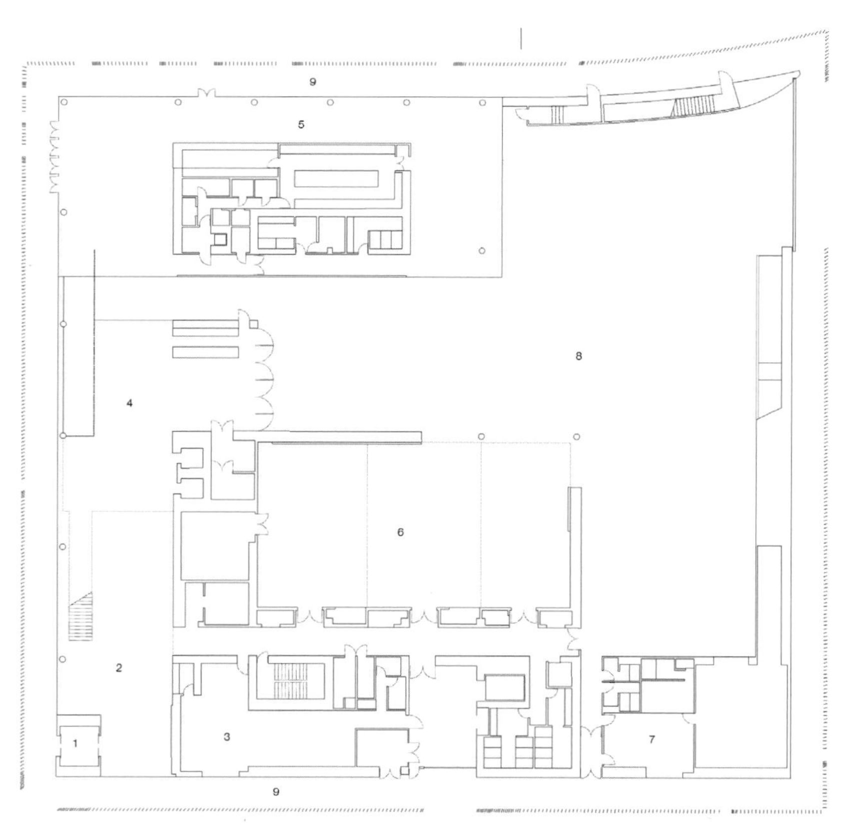

底层平面图 ground floor plan

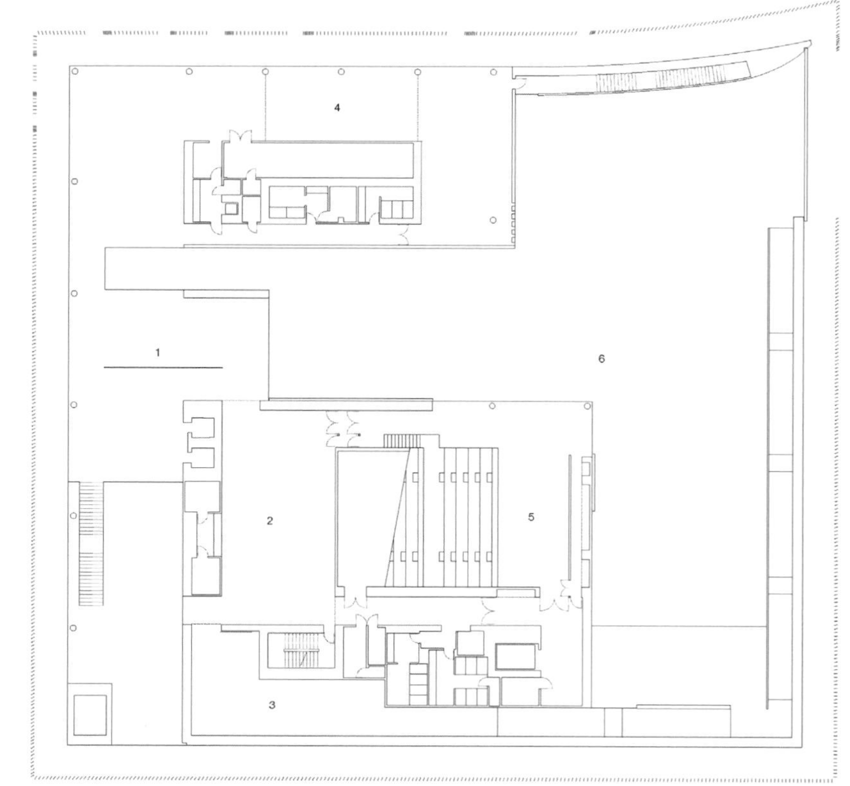

二层平面图 first floor plan

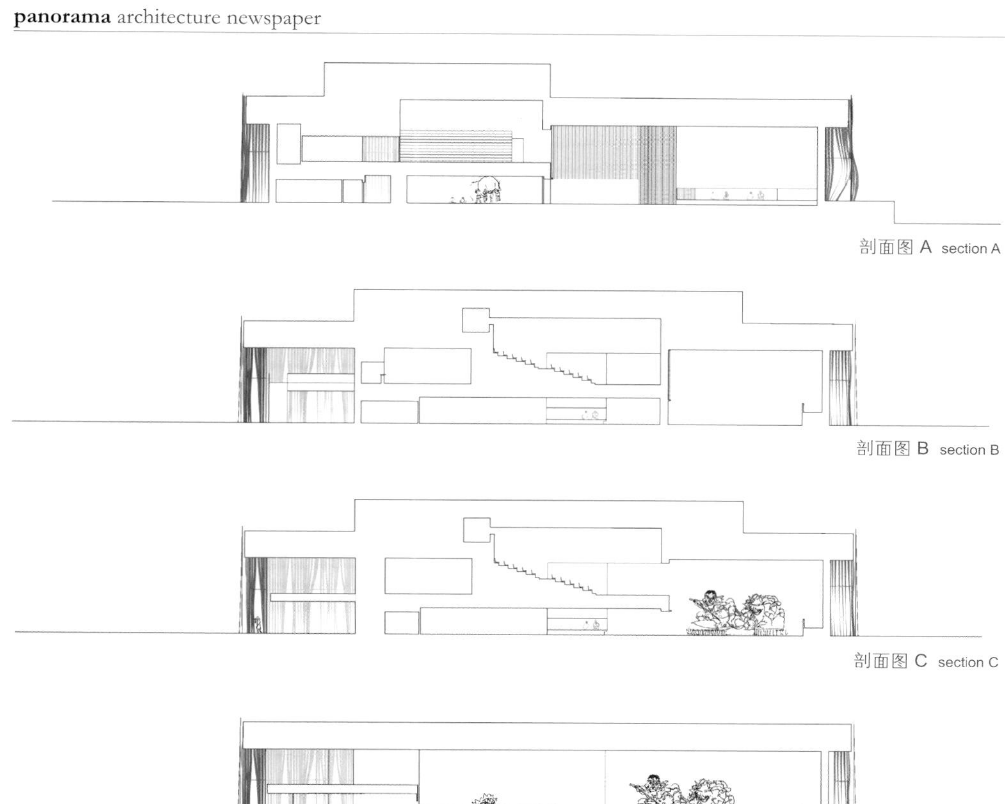

剖面图 A section A

剖面图 B section B

剖面图 C section C

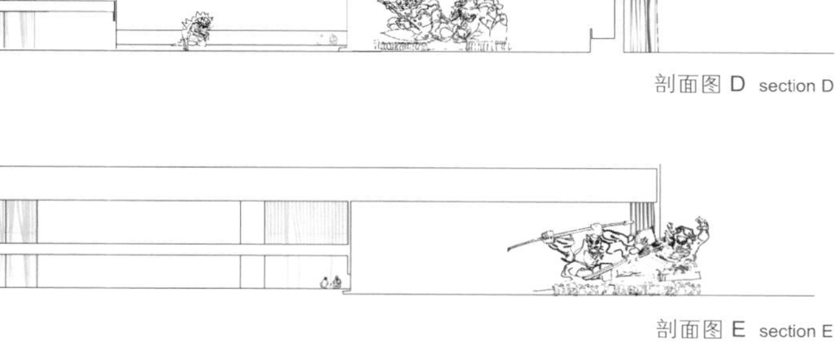

剖面图 D section D

剖面图 E section E

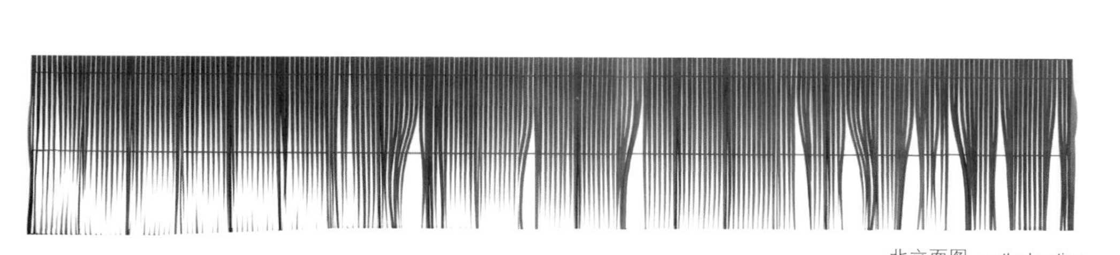

北立面图 north elevation

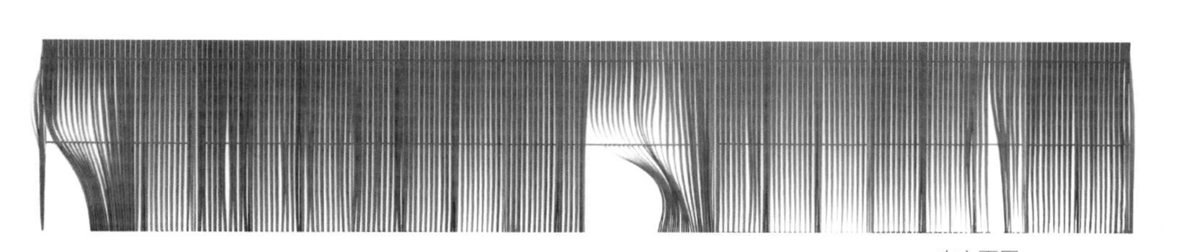

南立面图 south elevation

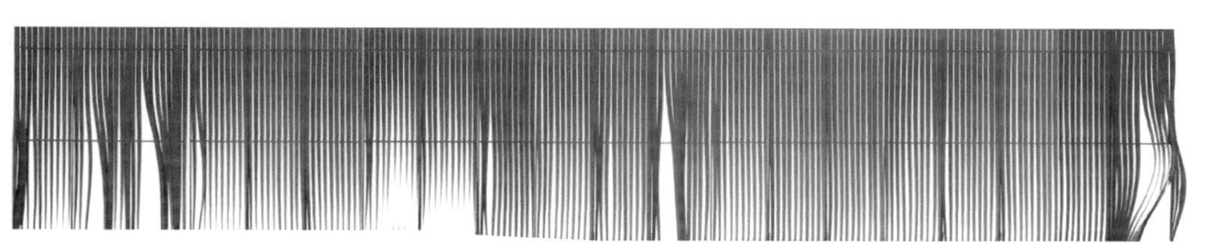

西立面图 west elevation

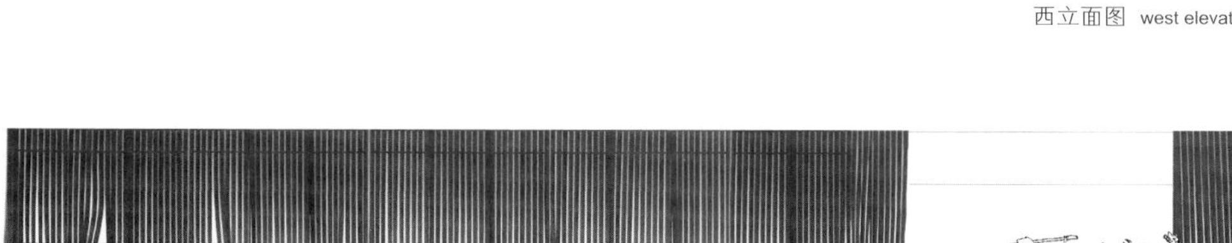

东立面图 east elevation

© Nakanimamasakhlisi Photo Lab

两个独立的玻璃体块
Two independent glass volumes

The building consists of two fifty meters by twenty meters horizontal planes, connected with ten peripheral columns and two independent glass volumes. The top horizontal plane acts as a common roof for both volumes. Lower horizontal plane acts as common platform and a public space.The building has received Best Project Award III at Moscow Architecture Biennale 2012.

Architects of Invention
"House of Justice",
Ozurgeti, Georgia
"司法部"，奥祖尔盖蒂，格鲁吉亚

该建筑由两块50m×20m的水平面组成，两个平面之间由十根檐柱和两个独立的玻璃体块连接。顶部平面作为两个玻璃体块的公共屋顶，底部平面作为公用平台和公共空间。该建筑在"2012莫斯科建筑双年展"上被评为"最佳项目奖第三名"。

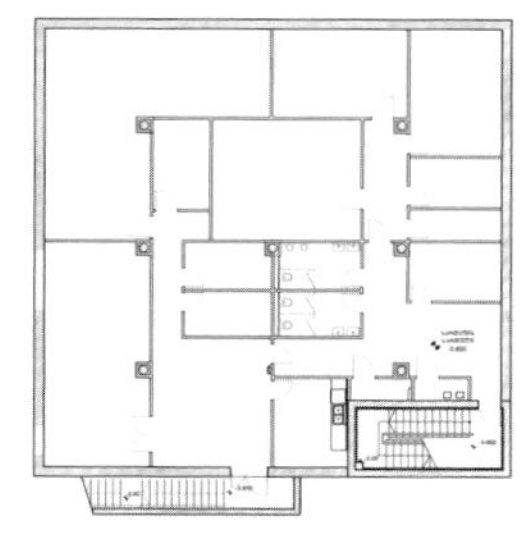

地下一层平面图 basement plan

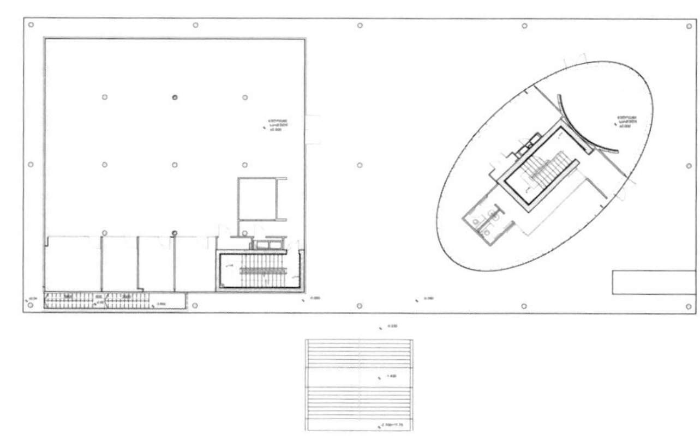

底层平面图 ground floor plan

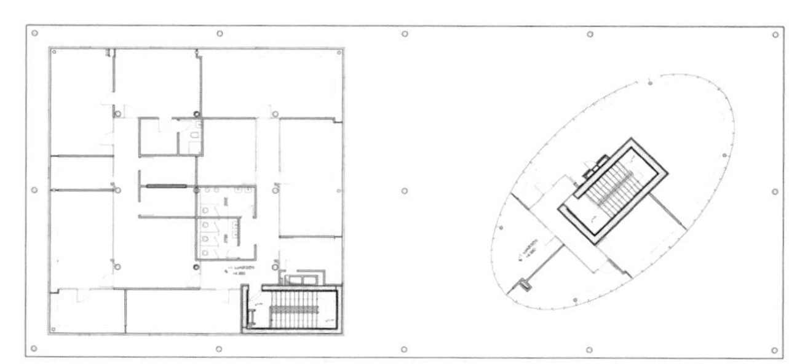

二层平面图 first floor plan

© Romain Ghomari (www.romwan.com)

一次设计演练

An exercise of reflection

DCA / Design Crew for Architecture
"Concept Stadium", France
"概念体育场"，法国

概念体育场项目是为法国国家橄榄球队设计主场体育场的一次设计演练，旨在提出橄榄球体育场的理想模式。如今，体育场被设计成一种自由物体，并且采用具有象征意义的形状从周围环境中脱颖而出。该项目为建造符合城市要求的实用型体育馆提出了一个建议。

The project of Concept Stadium is an exercise to think about a stadium to house French national rugby team, to propose what could be the ideal rugby stadium. Today, stadium is of iconic shape as autonomous object, distinguished from its environment. This is a proposal for a stadium to become a useful object with urban concerns.

该体育馆每天都将被使用

It will be daily used

传统环形体育场设施布置方案被设计成一条围绕在看台周围的连续坡道，为观众提供一条连续的通道。这条坡道被当做一条走廊，里面分布着许多商店和其他一些小型设施。

The traditional ring device distribution of the stadium is distorted to become a continuous ramp that circles the grandstands to offer a continuous fluid path. This ramp is treated as a corridor that could house shops and small facilities.

© Romain Ghomari (www.romwan.com)

一种激进的新型城市模型
A radical new urban model

MVRDV
"Urban plan of Almere Osterwold", The Netherlands
"阿尔梅尔·伍斯特伍德城市规划"，荷兰

这项开发战略旨在实现城市的有机发展，有效发挥人们的积极性，而且居住者能够创造他们自己周边的空间环境，如公共绿化、城市农业以及道路交通等设施。为了实现该城市的可持续性发展目标，本地区50%的区域将用于城市农业，在局部范围内为城市供应农产品，同时保持当前的农业特点。

The development strategy invites organic urban growth in which initiatives are stimulated and inhabitants can create their own neighborhoods including public greening space, urban agriculture and roads. To contribute to the city's sustainability goals, 50% of the area will be used as urban agriculture, producing products for the city on a local scale and maintaining the current agricultural character.

这是更加灵活的、完全以用户为导向的发展战略

It is completely user-oriented, more flexible

除了设计一份规划图，人们还必须实现道路交通、能源、卫生、废物收集、公共绿化以及城市畜牧等所有必需的城市功能。

Apart from designing a plan, one can achieve all necessary functions such as traffic, energy, sanitation, rubbish collection, public greening and urban farming.

中联筑境建筑设计有限公司 CCTN.ARCH
"Xiasha Scientific and Cultural Center", Hangzhou, Zhejiang, China
"下沙科文中心"，杭州，浙江，中国

项目用地位于城市核心区内，金沙湖北畔，与市民中心相邻。项目由大剧院和科文中心组成。大剧院整体呈现一种动感轻盈的城市形态，柔性轮廓沿金沙湖和城市广场扭曲并抬升，犹如柔软的杭州绸缎一般，向公众传达一种亲切的欢迎姿态。科文中心建筑整体呈U形布局，向南侧的文化广场及金沙湖景观打开。同时绿意盎然的"文化客厅"向南延伸至文化广场，一气呵成，并与环抱状的建筑实体形成了阴阳相抱、合二为一的有机组成关系。

The project is located on the north bank of Jinsha Lake, a urban core area, adjacent to the Citizen Center. The project consists of Grand Theater, Scientific and Cultural Center. The Grand Theater adopts a dynamic and lively urban shape whose soft silhouette is twisted and lifted up along the Jinsha Lake and the urban square, resembling a piece of soft Hangzhou silk fabric and making a welcoming gesture to the public. The Scientific and Cultural Center adopts a U-shaped layout, opening up to the cultural square and Jinsha Lake at south. Meanwhile, the green "cultural parlor" is connected to the cultural square at south and creates an integrated organic relationship with the circular building body.

精巧柔和的轮廓
Soft and subtle silhouette

紧凑型建筑
A compact building

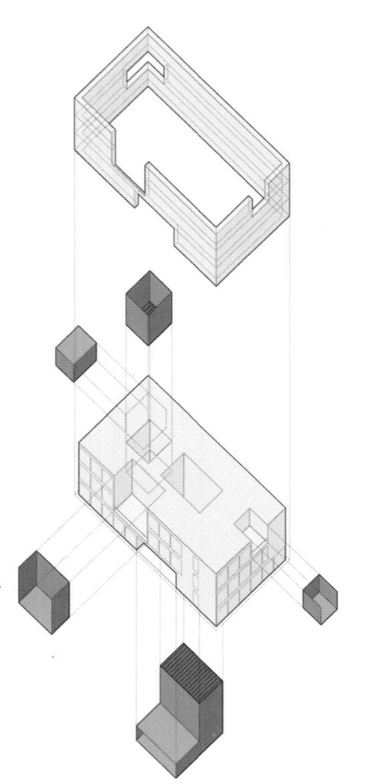
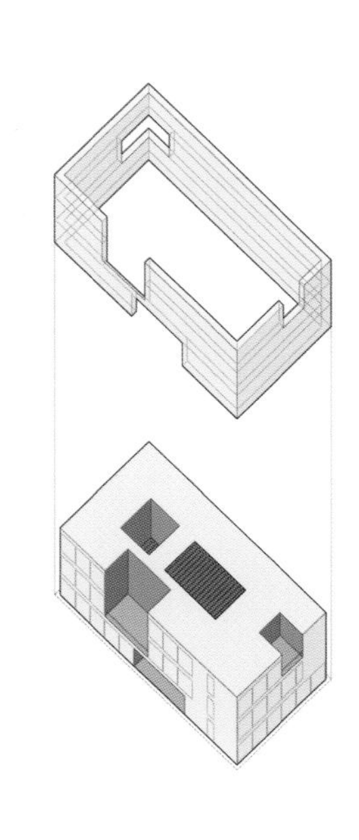
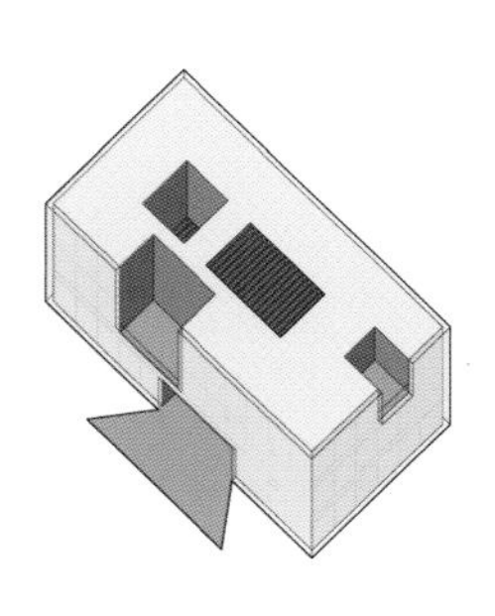

Dosmasuno Arquitectos (Ignacio Borrego, Néstor Montenegro, Lina Toro)
"Social Services Centre", Mostoles, Madrid, Spain
"社会服务中心"，莫斯托雷斯，马德里，西班牙

该建筑项目需要考虑两个主要方面。一方面，该建筑拥有许多结构完全相同的单人办公室，并且位于一个正在建设的新无菌城市区域内，该区域内除了街道网和地块朝向外没有任何限制因素。另一方面，基于对环境的尊重，该建筑与附近和远处环境元素之间的关系通过传统方法得以实现。

热量与视觉过滤器把该建筑从外部空间有效地隐藏起来

该项目一期包括减轻质量、压缩体积以及创造外部等候室或工人休息区等开放式户外空间。二期包括插入一个单一体块和一个包括主要通道和多功能大厅的连续空间。最后阶段为该建筑包裹一层薄薄的外皮。

The programme of the building is strongly conditioned by two facts.On the one hand, it hosts many identical spaces — single offices — and it is located within a new, aseptic urban fabric — currently under construction — which features no constraints but the street network and the plot orientation. On the other hand, the building's relationship with its close and distant surroundings is based upon environmental respect, achieved by means of traditional procedures.

A thermal and visual filter conceals the building from the outside

The first operation consisted of lightening the volume through the extraction of mass, generating open, exterior spaces as external waiting rooms or resting areas for the workers. The second operation consisted of inserting a singular volume, a continuous space, which hosts the main access and the multi-purpose hall. Finally the building is wrapped by a thin skin.

© Miguel de Guzmán www.imagensubliminal.com

© Miguel de Guzmán www.imagensubliminal.com

© Rubén Pérez Bescós

© Rubén Pérez Bescós

一种创新模式
An innovative pattern

Vaillo+Irigaray Architects
"CIB Biomedical Research Center", Pamplona, Spain
"CIB生物医学研究中心"，潘普洛纳，西班牙

该建筑结构外面包裹的铝壳具有很强的视觉吸引力，内部空间包括医院和研究设施。外部表皮延续这个大型医学建筑的室内空间特点。

The aluminum envelope stands out as a visual focal point. The interior houses a hospital and research facility. The outer skin traces the internal features of the large medical structure.

受自然界启发的立面系统

建筑立面类似锯齿状刀刃的边缘，保护室内空间免受强烈的太阳辐射，并能有效调节室内温度。

A façade system inspired by nature

The façade, which resembles the edges of a serrated blade, shades the interior from sun radiation while regulating its temperature.

© Rubén Pérez Bescós

© Rubén Pérez Bescós

© Rubén Pérez Bescós

© Rubén Pérez Bescós

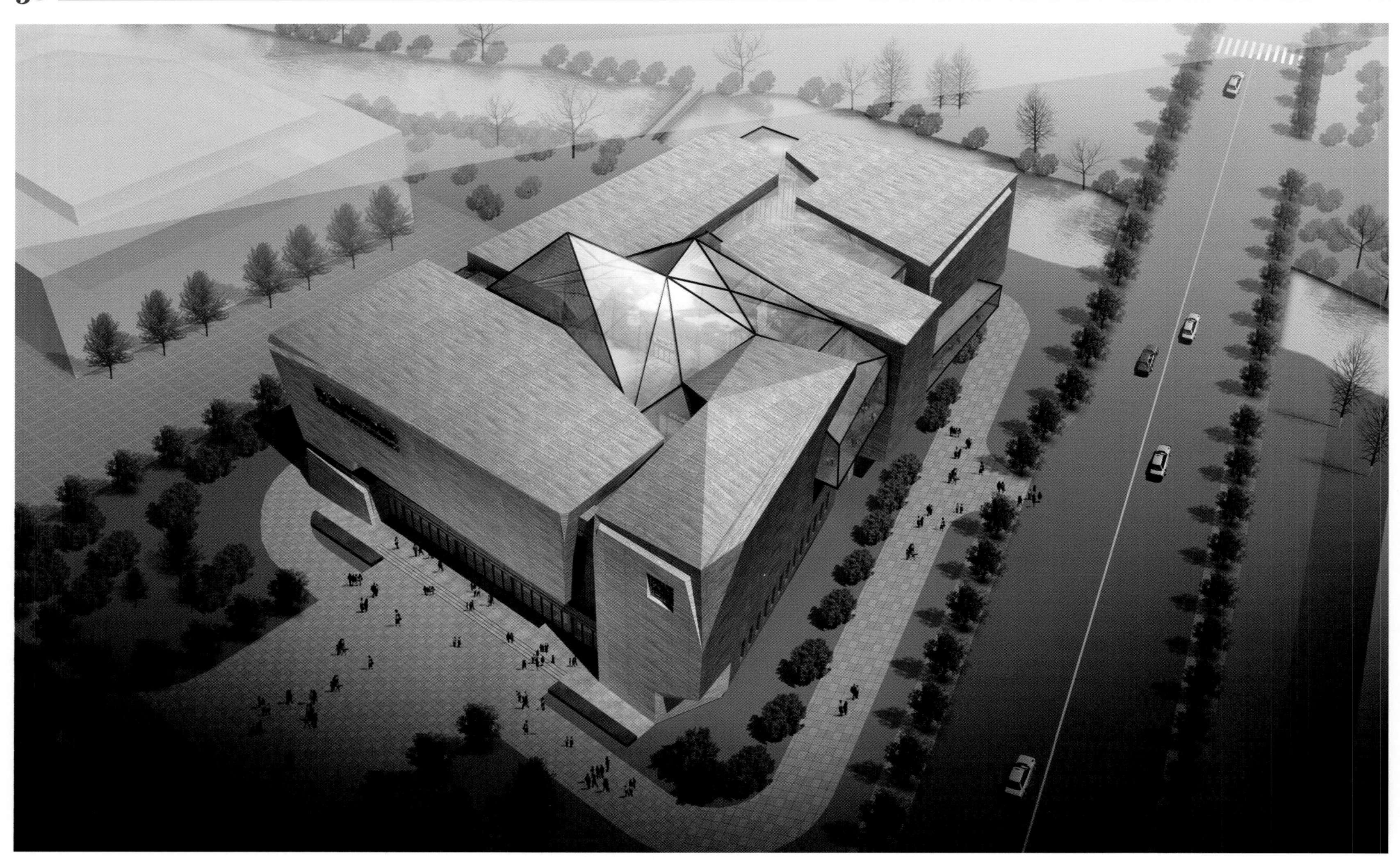

朴实而永恒的外观
A secular objective manner

中联筑境建筑设计有限公司
CCTN.ARCH
"Huangyan Museum ",
Huangyan, Taizhou, China
"黄岩博物馆"，黄岩，台州，
中国

项目位于浙江省台州市黄岩区南面，总用地面积9 809m²，地上建筑面积约10 000m²，地下1层，地上3层，属以收藏展示功能为主的、专业性较强的大型博物馆。

设计以黄岩的石文化为出发点，创造坚固、稳重的博物馆形象。博物馆造型犹如五块巨石，稳稳地坐落于基地之上。顶上有玻璃体块穿插其中，如黄岩溪水流淌其间。展陈功能使用人工采光，建筑立面以实墙面为主，也符合建筑本身石文化的意向。从大厅行至中庭，沿扶梯而上，上空廊桥飞架，参观者仿佛穿梭于巨石之间，空间迂回盘旋，石梁横卧，让人想到黄岩的锦绣胜景。

This project is located at south of Huangyan District, Taizhou, covering a total land area of 9,809 m^2. It consists of one floor underground and three floors above ground, providing about 10,000m^2 building area. It is a highly-dedicated large museum mainly used for collection and exhibition.

The design is based on the stone culture of Huangyan District, creating a solid and stable museum identity. The museum resembles five pieces of huge stone stably standing on the base. Glass volumes are inserted on the top, like water flowing through. Artificial lighting is adopted for exhibition function and most façades are solid walls responding to the stone culture concept of the building. Walking from lobby to atrium and ascending along the stairway, visitors will see overhead gallery bridge providing an experience of passing through huge rocks, and circuitous spaces and transverse rock beams evoke brilliant and beautiful scenes of Huangyan.

绿色城市皮肤

Broken volumes

中联筑境建筑设计有限公司
CCTN.ARCH
"Xiangtan Urban Planning Exhibition Hall and Museum", Xiangtan, Hu'nan, China
"湘潭城市规划展览馆及博物馆"，湘潭，湖南，中国

建筑位于城市新区核心地带，紧邻行政中心和新区梦泽湖景观，交通便捷，环境优美。其主要由博物馆、规划展示馆、规划局办公等功能组成。设计理念立足于湘潭的地理环境与人文历史，红色的基座、灰白的建筑主体由黑色的构架连为一体，整体寓意着湘潭"格物致知"的人文底蕴与"经天纬地"的雄才大略，同时也体现了建筑的时代性。

The building is located at the core area in new urban district and is adjacent to administrative center and Mengze Lake, enjoying convenient traffic and beautiful environment. It consists of the following main functions, museum, planning exhibition hall and planning bureau offices, etc. The design concept is based on geographic environment and humanism history of Xiangtan. Red base and offwhite main body are integrated through the black truss, embodying the humanism culture of "study the nature of things" and great ambition of "planning the whole nation", and simultaneously representing the contemporary characteristics of the building.

水平棒体
Horizontal bars

BIG
"Cross # Towers", Seoul, Korea
"交叉#字塔楼"，首尔，韩国

这个2.1万平方米的塔楼设计方案紧邻MVRDV设计的连续云端项目以及由KPF建筑师事务所、蓝天组建筑事务所(Coop Himmelblau)、REX建筑事务所、山本理显设计工场(Riken Yamamoto Fieldshop)和伦佐·皮亚诺在Libeskind建筑事务所开发的Yongsan国际商业区项目中的设计方案。

在水平和垂直方向相互交叉的三维城市社区塔楼

为了给该项目提供足够的广场空间并满足建筑高度的要求，交叉#字塔楼的部分结构相互交叉，形成相互连接的天桥。这些水平塔楼同时也作为绿化屋顶空中花园的基础。

The 21,000 m² tower proposal will sit alongside the contentious cloud project by MVRDV as well as schemes by KPF, Coop Himmelblau, REX, Riken Yamamoto Fieldshop and Renzo Piano within Yongsan masterplan developed by Studio Libeskind.

A three-dimensional urban community of interlocking horizontal and vertical towers

To provide enough square footage for the project and meet building height requirements, part of the Cross # Towers building is turned it on its side to create interlocking bridges. These horizontal towers serve as the foundation for green-roofed sky gardens.

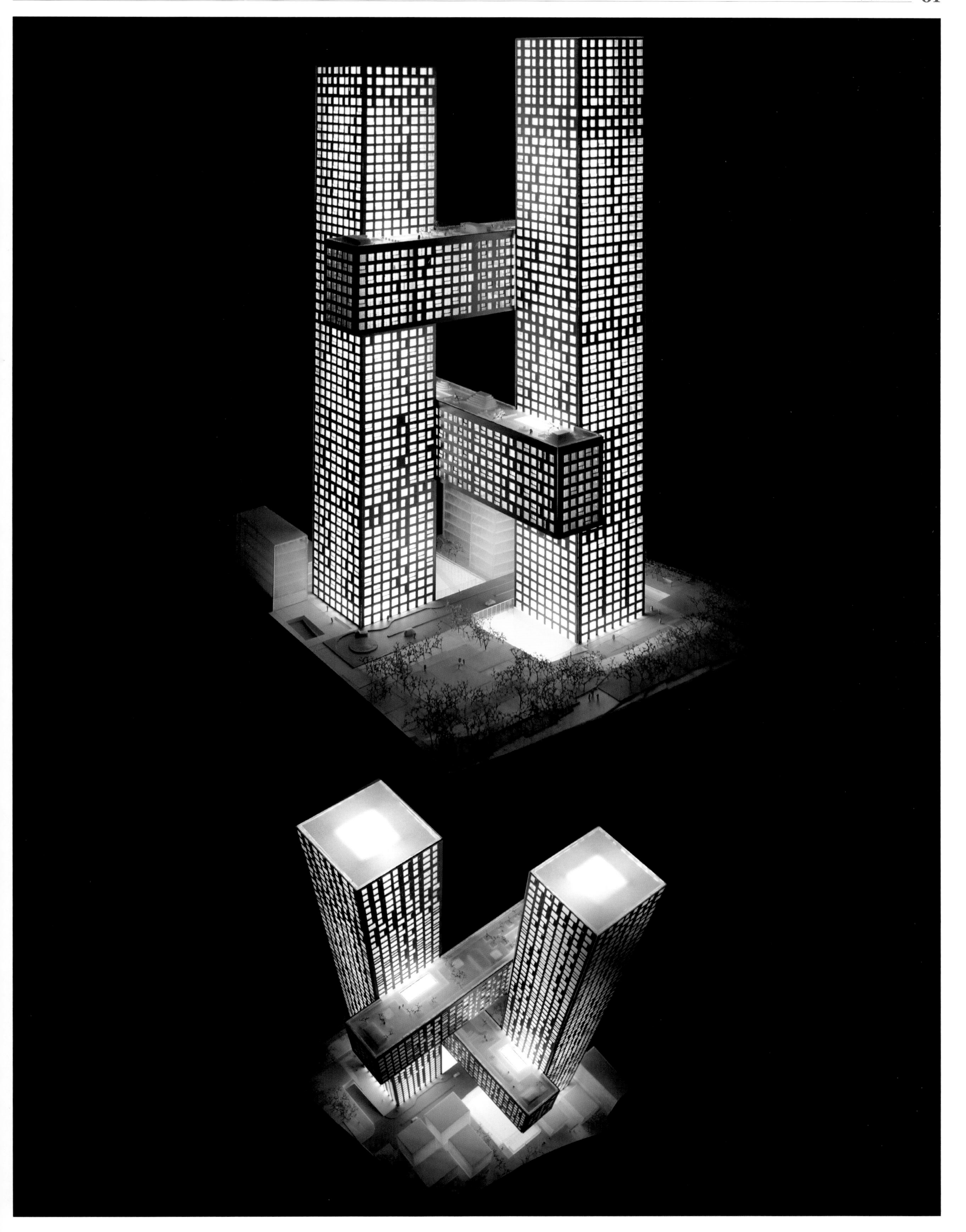

住宅
RESIDENTIAL

© AQSO arquitectos office

© AQSO arquitectos office

连续体块

A continuous volume

AQSO
"Residential Complex",
Casablanca, Morocco
"住宅综合体",
卡萨布兰卡，摩洛哥

设计方案将项目地块沿街道规划成一个连续建筑，中间围绕着两个大型传统住宅建筑。

建筑高度根据场地内环境条件而改变，将私人住宅和城市环境巧妙地结合到一起

建筑立面采用两种不同的设计风格：朝向公众环境的外墙具有内向和严肃的特点；而朝向私人庭院的内立面具有外向和朴实的特点。

The scheme is conceptually solved as a continuous block aligned with the surrounding streets and wrapping around two big 'riads'.

Its height adapts to the different conditions of the plot to combine a domestic and urban appearance

The building façade turns into two different strategies: the exterior skin facing the most public context becomes an introverted and formal element while the interior one facing the private courtyards becomes extrovert and domestic.

© AQSO arquitectos office

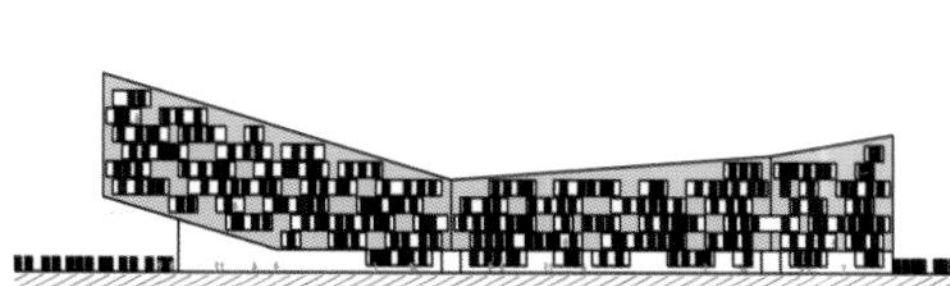
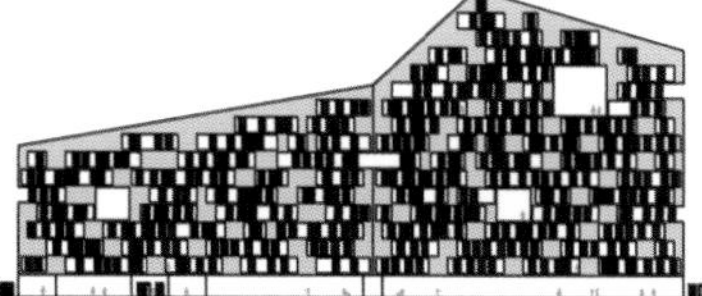
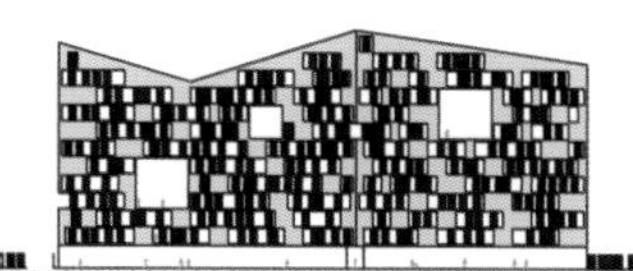
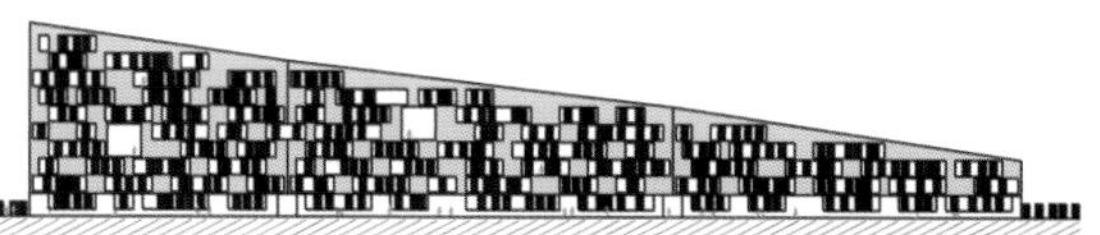
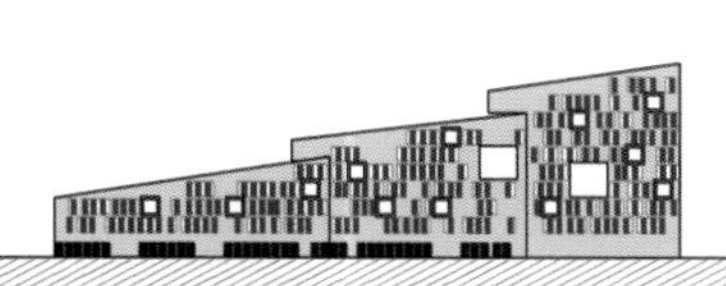
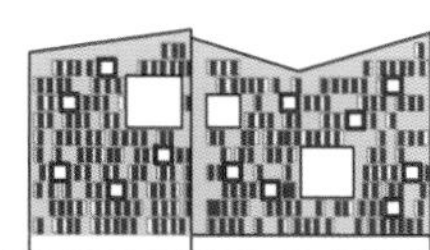
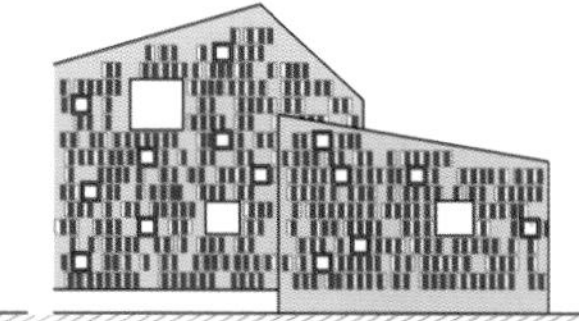
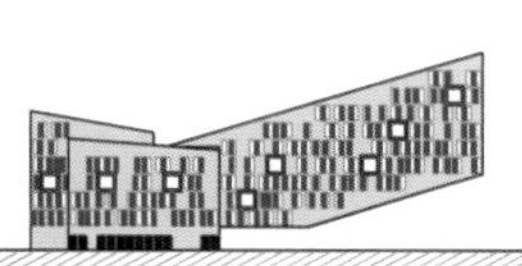

立面图 elevations

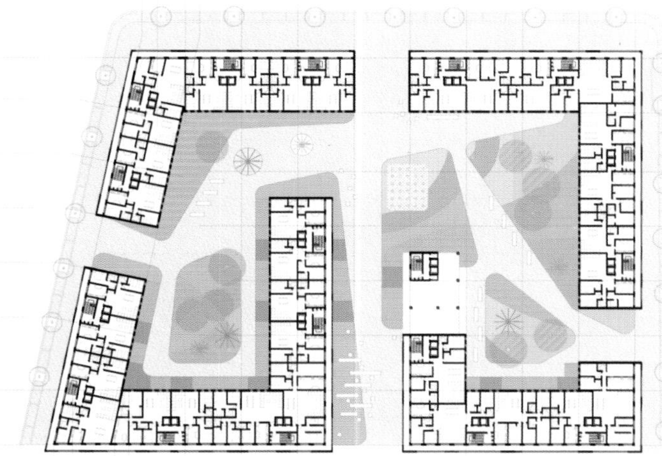

楼层平面图 2 floor plan 2

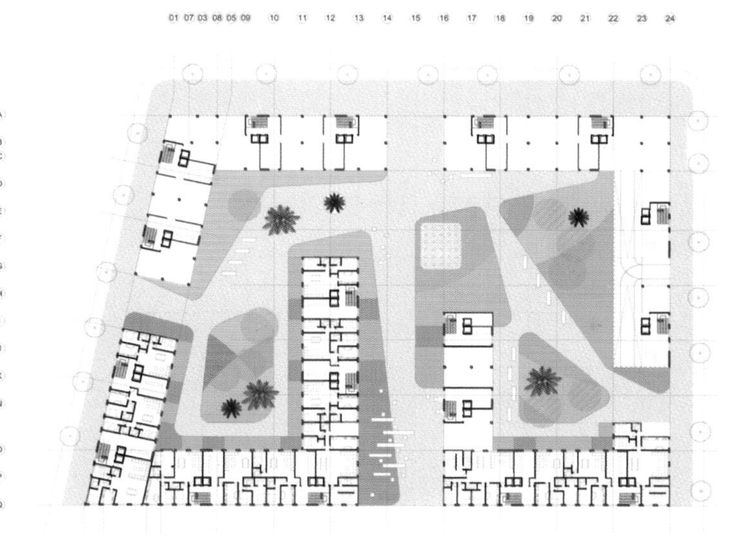

楼层平面图 1 floor plan 1

Long Studio © Bent René Synnevag

Saunders Architecture
"Six Studios", Fogo Island, Newfoundland, Canada
"六间工作室"，福戈岛，纽芬兰，加拿大

福戈岛(Fogo Island)位于加拿大纽芬兰东北海岸大约12英里处，人口总数为2 500人。在最近几十年内，福戈岛上的居民亲眼目睹了鳕鱼渔业的萧条和缺乏支撑当地经济发展的金融投资给他们的传统生活方式带来的影响。

这六间工作室是独立的建筑系统，不依赖市政供水、排污、天然气、供电或类似的公共服务业

分布在福戈岛上的这六块偏远地块将用于实施艺术中心负责的部分设计项目。这六间工作室将供艺术家和作家使用，这样他们就可以每天与当地居民进行交流，体验各种社区生活。其中第一间工作室，即长形工作室，于2010年6月竣工；桥形工作室、塔形工作室和扁形工作室于2011年6月正式开放；其余工作室将于2012年竣工。这六间工作室的设计意图是为当地施工技术提供样板与参考。

Long Studio © Bent René Synnevag

Squish Studio © Bent René Synnevag

Fogo Island, located about twelve miles off the northeast coast of Newfoundland, Canada, supports a population of twenty-five hundred people. In the last several decades, the people of Fogo Island have seen their traditional way of life been challenged by a diminished cod fishery and lack of financial investment to sustain the local economy.

All six studios are autonomous — they do not rely on municipal water supply, sewer, natural gas, electrical power grid or similar utility services

Six remote sites scattered across the Fogo Island were chosen to host a portion of the art centre's programs. The six studios for artists and writers, allows them to live within the various communities and interact, on a daily basis, with the local residents. The first of these studios, the Long Studio, was completed in June 2010. The Bridge Studio, Tower Studio and Squish Studio were officially opened in June 2011. The rest will be completed in 2012. In all six studios, the intent was to sample and allude to local construction techniques.

Bridge Studio © Bent René Synnevag

Tower Studio © Bent René Synnevag

Tower Studio © Bent René Synnevag

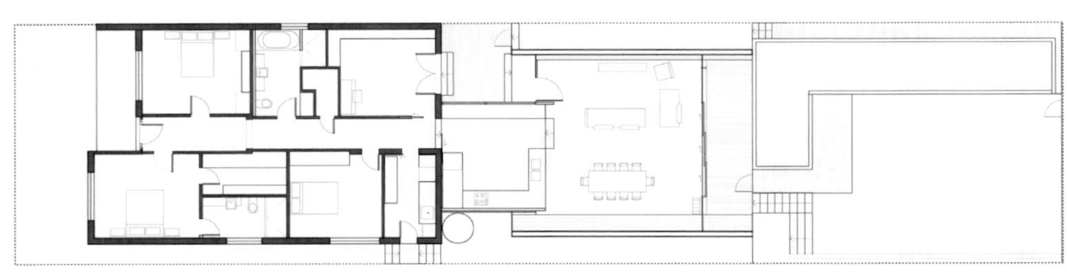

平面图 plan

Mcbride Charles Ryan
"Cloud House",
Melbourne, Australia
"云朵住宅"，墨尔本，
澳大利亚

"云朵住宅"项目包括对一个爱德华时代双正面住宅进行扩建和翻新。该住宅包括三部分，当从中穿越时，你能够瞥见前面即将出现的空间和体验。

凸出的云状空间是出人意料的边缘空间

原建筑结构中的空间大部分采用白色，并配以有植物图案的门厅地毯。穿过门厅，一个分解式的红色"盒子"赫然出现在人们眼前。扩建的云朵状边缘空间给人们留下儿童时代的印象，家人和朋友可以在这里就餐或者在弧形区域内自由玩耍。

The Cloud House is an addition and renovation to a double-fronted Edwardian house. The three parts of the house allow for glimpses previewing oncoming spaces and experiences as you move through the home.

A cloud-shaped extrusion is the unexpected final space

The spaces within the original structure are largely white in colour, united by exotic floral hallway carpet. This journey through the space is followed by encountering a disintegrated red-coloured "box". Following the form of a child-like impression of a cloud it is a playful addition where family and friends can eat and have fun surrounded by the curved form.

独特的外观造型
Distinct and unexpected episodes

© John Gollings

© John Gollings

© John Gollings

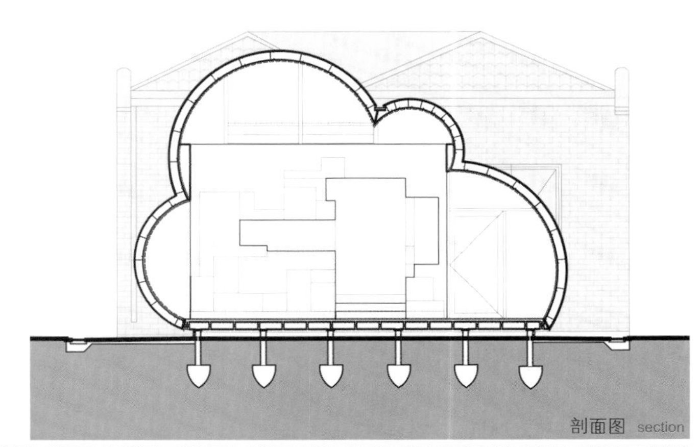

剖面图 section

宁静的空间
Quiet at last

MVN Arquitectos
"AA House",
Almeria, Spain
"AA制住宅",
阿尔梅里亚，西班牙

所有设计工作的核心是开发一些策略使该项目的核心区域能够朝向地平线打开。建筑师们确保地平线将成为该住宅的边界。

设计原则是适应场地地形

该住宅被划分为彼此纵向错位的三部分，从而创造出私密空间。

All the work was focused on developing project strategies that would enable to open the heart of the project towards the horizon. The architects ensured that the limit of the house should be the horizon.

It has been defined by the need to adapt to the topography

The house is organized into three bands that are displaced longitudinally and seeks for shelter and protection.

© Nelson Kon

两个截然不同的空间
底层空间由两排14根细钢柱组成

Two contrasting spaces
The lower level is comprised of two rows of 14 slender steel columns

Studio Mk27
"Toblerone House",
Sao Paulo, Brazil
"巧克力住宅"，圣保罗，巴西

简单的设计理念配以简洁的建筑结构，这个长形二层住宅被划分成两个不同的区域。一楼是社交和起居空间。当大块玻璃窗户打开之后，起居室就会变成一个毫无遮挡的开放空间，完全向户外花园敞开，使该项目成为一个支柱上的住宅。二楼采用可伸缩木板进行封闭，在形成更加私密空间的同时，为室内空间提供遮阳功能。

© Nelson Kon

The conceptual and programmatic simplicity of the house joins a structural simplicity. This elongated two-story house is divided into two distinct sectors. The ground floor is the social living space. When the massive glass windows are opening, the living room becomes a free space, totally opening to the gardens — a house on pilotis. The upper floor is more secluded as it is closed off by retractable wooden panels allowing more privacy and shade from the Brazilian sun.

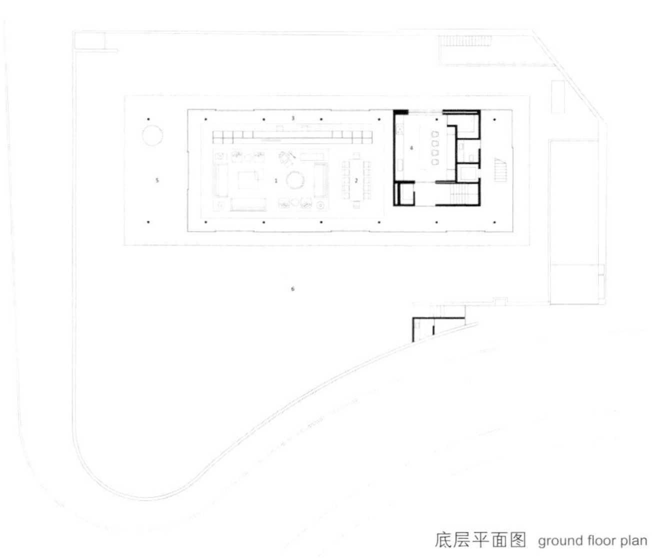

底层平面图 ground floor plan

上层平面图 upper floor plan

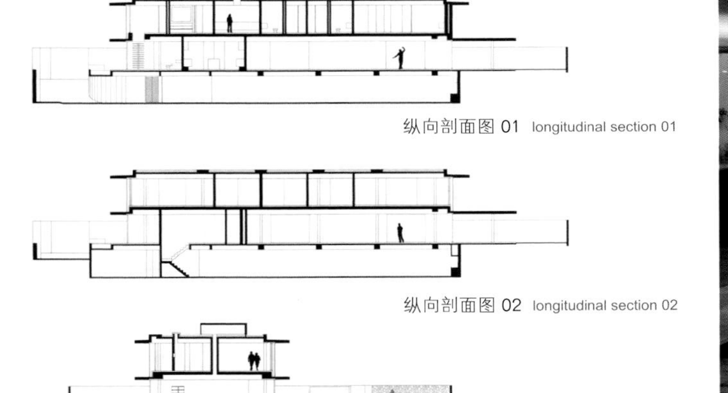

纵向剖面图 01 longitudinal section 01

纵向剖面图 02 longitudinal section 02

横向剖面图 cross section

© Nelson Kon

© Nelson Kon

© Nelson Kon

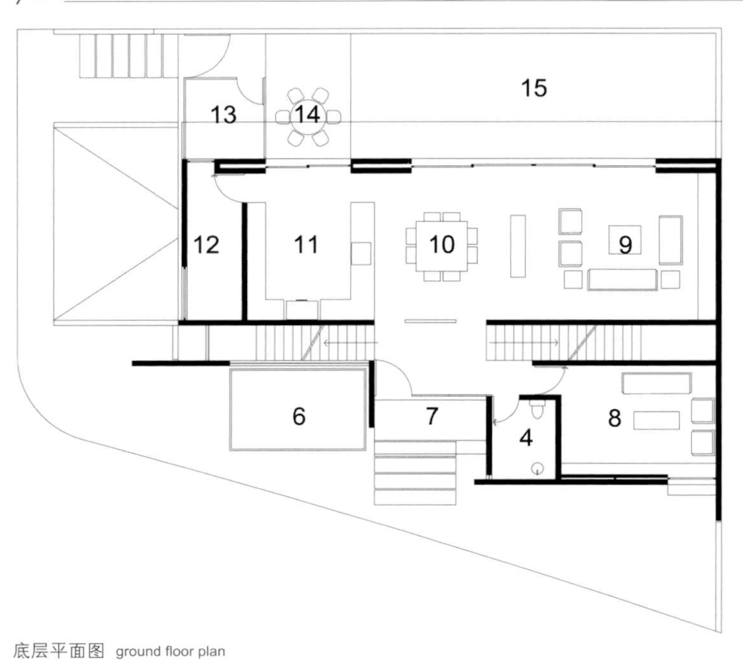

底层平面图 ground floor plan

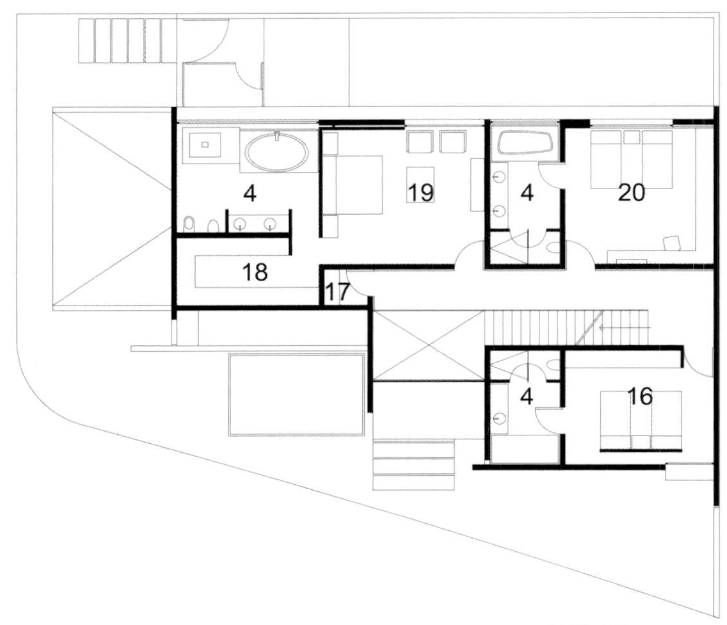

二层平面图 first floor plan

基础设计图
A basic diagram

Agraz Arquitectos
"X House", Jalisco, Mexico
"X住宅"，哈利斯科，墨西哥

该住宅位于场地的一角，毗邻一个已建成项目，使其能够与附近住宅进行对话。

建筑由白色混凝土主体、大理石面板和金属网格组成

面板之间的虚实搭配形成一系列动态空间，细板条在实现形式表达功能的同时还能将光线引入室内。

The house is situated on a corner site, next door to a previously built project which created the opportunity to establish a dialogue with the neighboring dwelling.

It is a combination of white concrete bodies, marble plates and metal lattice work

The proximity and relationships of the planes form a series of dynamic spaces through a play of solids and voids, while thin slats shape the formal expression and allow light inside.

© Mito Covarrubias

倾斜的墙面
Sloping walls

Keitaro Muto Architects
"Ginan House",
Hashima-gun, Gifu, Japan
"岐南住宅"，羽岛郡，
岐阜，日本

该住宅的外观使其看上去好像花园的一部分。建筑外墙采用砂砾涂层进行装饰，其中一面墙体向外倾斜，形成一个楔形体块。

由多种几何形状组成

建筑立面的凹进处设有一个小型泳池，将组成该建筑的两个相互重叠的体块彼此分开。

The façade make the building like a part of the garden. Gravel coats the exterior of the house, including a wedge-shaped block with outward sloping walls.

It blends varied geometric shapes

A small swimming pool is tucked into a recess in the façade, marking the division between the two overlapping blocks that comprise the building.

© Yoshiike Teruaki

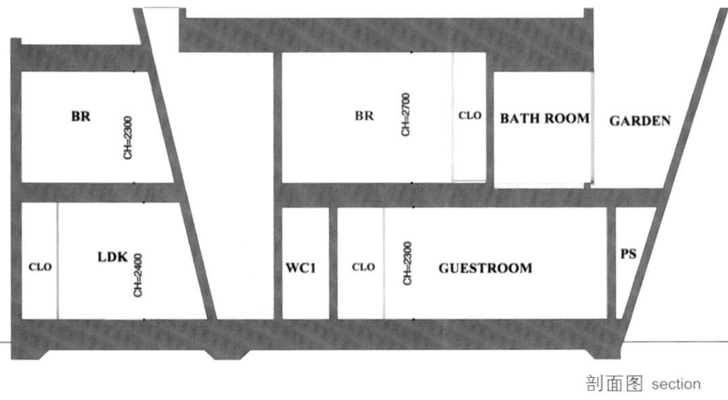

剖面图 section

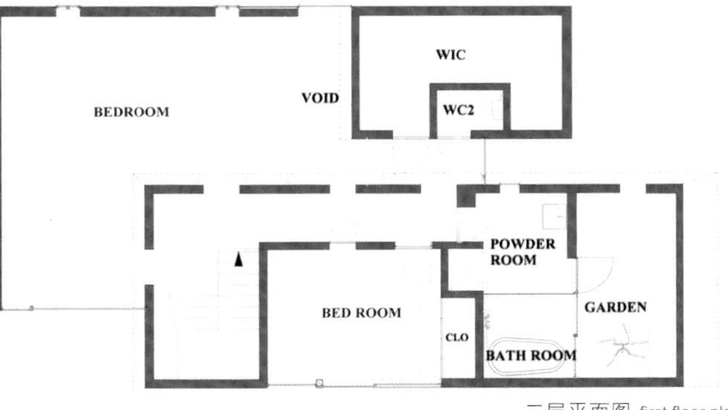

二层平面图 first floor plan

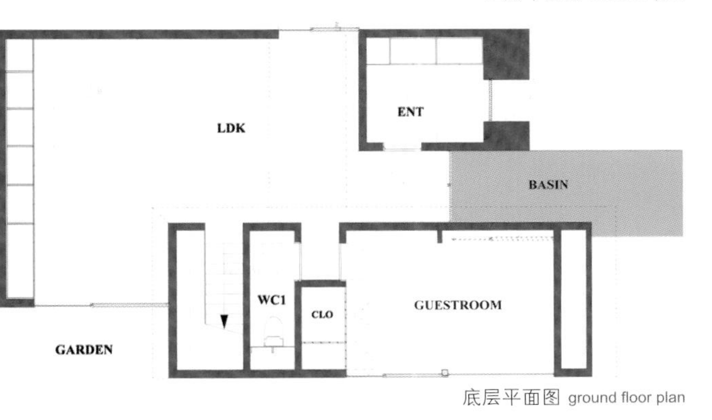

底层平面图 ground floor plan

© Yoshiike Teruaki

剖面图 sections

碉堡一样的住宅
Like a fortress

Formwerkz Architects
"Armadillo House", Singapore
"犰狳住宅"，新加坡

该设计理念必须考虑以下三个主要因素：来自附近公路的噪声污染、沿建筑侧面照射的刺眼阳光以及室内空间的私密性。此外，该住宅设计还要具有灵活性和可持续性，以便满足家庭成员不断增加的要求。

既要向室外环境敞开，又要保持最小的噪声影响，这一点是能够实现的

解决方法是采用四块弯曲的外壳保护外部建筑元素，阻挡直射阳光，同时保护内部空间的私密性。

© Jeremy San

The concept had to respond to three main factors: the noise pollution from the adjacent highway, the hard western sun beating down along the broad side of the dwelling and the privacy. Moreover the house had to be flexible and sustainable in its design for a growing family.

It is possible to open to the environment with the most minimum noise impact

The solution is four stepped bent shells that block out the external elements and direct light and protect the intimate interior.

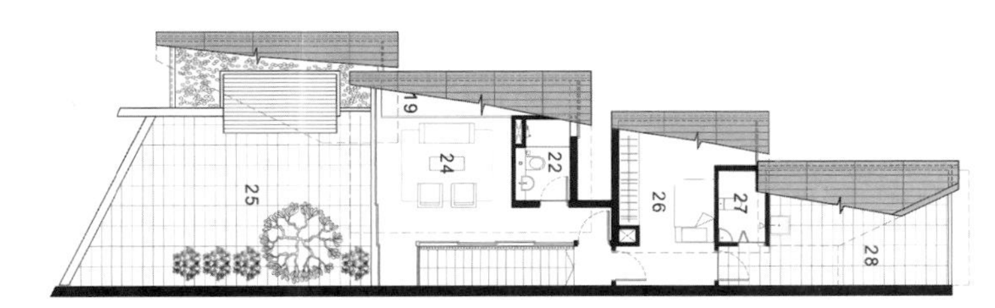

三层平面图 second floor plan

二层平面图 first floor plan

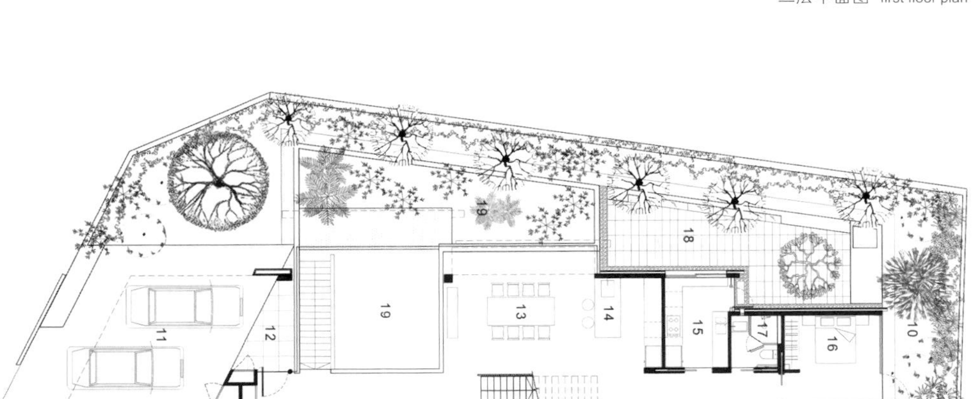

底层平面图 ground floor plan

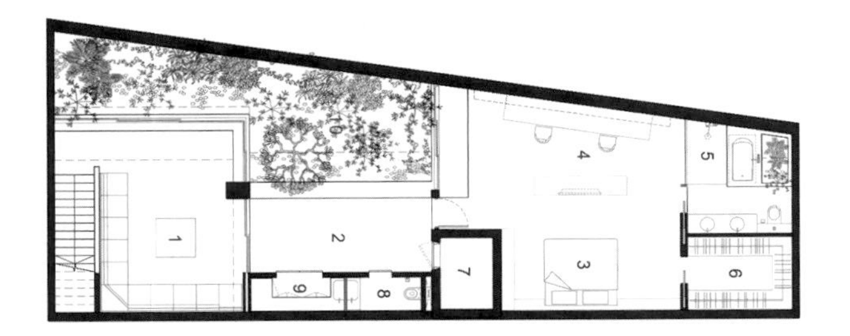

地下一层平面图 basement plan

© Stéphane Chalmeau

阳台采用木质与金属混合材料
Wood and metal balconies

Peripheriques Architectes + Berranger & Vincent Architectes
"72 Apartments Complex", Nantes, France
"72套公寓综合体"，南特，法国

该建筑围绕一个南向中央花园而建，阳光从中央花园的缺口处能够直接到达该地块的中央区域，从而为优化公寓的朝向和景观视图提供了良好的条件。

The building was built around a southern-oriented central garden which allows the sunlight to penetrate deeply in the heart of the block, and therefore offers good opportunities to optimize the views and orientation of the apartments.

建筑外观的颜色根据亮度呈现出多变的效果

The colour of the building façades changes depending on the luminosity

建筑最高处为9层，最低处为3层。整个建筑为单一体块，漆面钢板制成的"鳞片"状灰色阴影覆盖整个建筑立面。

The volume reaches its pinnacle up to 9 levels, before coming down again to the 3rd floor. The monolithic volume is characterized by grey-coloured shaded "scales", made of lacquered steel, which cover the façades.

© Stéphane Chalmeau

平面图 +2 floor plan +2

© Stéphane Chalmeau

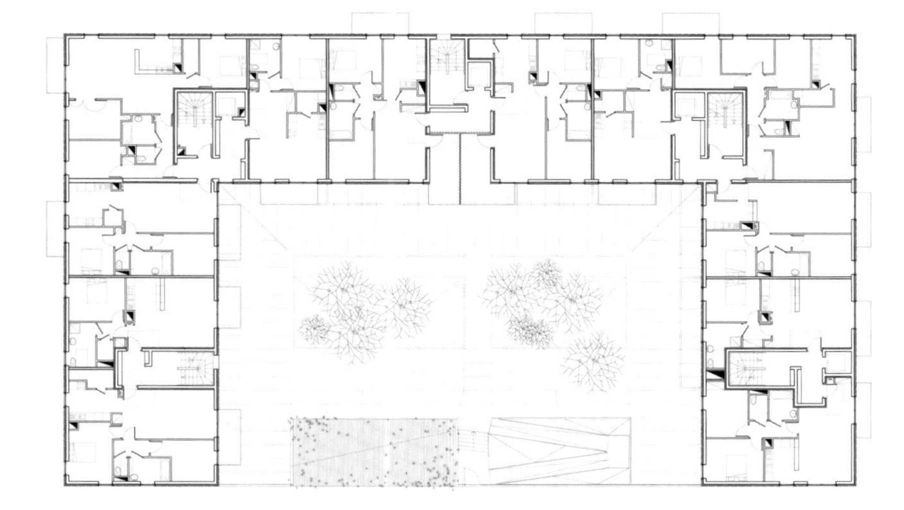

平面图 +1 floor plan +1

© Stéphane Chalmeau

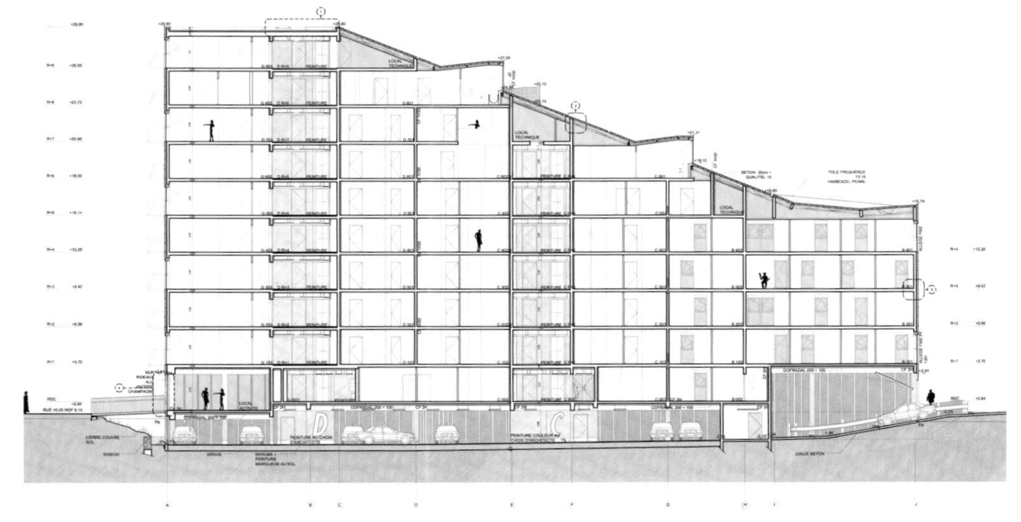

剖面图 section

© Stéphane Chalmeau

立面图 elevation

© Stéphane Chalmeau

设计
DESIGN

漂浮在空中的古老宝石
Levitating historic gems

Tom Dixon
"Lustre Pendants"
"闪亮的垂饰"

该设计使人们联想到孔雀羽毛或水面上的浮油等华丽的颜色变化效果。手工制作工艺使每个物体具有唯一的、不可复制性的装饰效果。

The result is a striking colour change effect reminiscent of peacock feathers or oil slicks on water. The handmade process results in a totally unique and unrepeatable finish for each individual object.

© James Field

眼球座椅
Eyeballs chair

Fiona Roberts
"Scopophilia"
"窥视狂"

澳大利亚设计师菲奥纳·罗伯茨(Fiona Roberts)设计了一款在鲜艳的红色天鹅绒座面上镶嵌了300颗塑料眼球的座椅。

Australian designer, Fiona Roberts, has created a chair upholstered in lush red velvet with 300 plastic eyeballs.

一种痴狂的艺术

A delicious slice of paranoid art

这款座椅是罗伯茨在澳大利亚阿德莱德市赫尔普曼学院(Helpmann Academy)2012年毕业设计展上的毕业设计。

The chair sculpture was part of Robert's graduate show at the Helpmann Academy Graduate Exhibition 2012 in Adelaide, Australia.

更加接近大自然
Closer to nature

Goodss Passion
"GO Recycle Bin"
"GO垃圾桶"

我们正处在一个大规模生产、处理和回收的年代。对于大多数人来说，回收利用是一件令人头痛的事情。

We are at the age of mass production, mass disposal, and mass recycling. For most people, recycling is boring.

我们是否需要一些动力对物品进行回收利用?

Do we need motivation to recycle?

通过创新性和创造性的设计，这些垃圾桶将改善家庭废物的循环利用过程，使家庭废物的循环利用更加简单、有趣和方便。一项具有影响力的优秀设计对于人们理解这个世界的真正需求是必不可少的元素。GO垃圾桶已经获得了多项设计奖，其中包括德国红点设计奖、香港创意奖、香港全球设计金奖以及香港最佳设计等。

Through innovative and creative design, these bins want to improve the process of household recycling by making it simple, fun and convenient. Good and influential design is essential for one to understand the real needs of this world. GO recycle bin has won numerous design awards including Germany's RedDot Design Awards, HK Perspective Award and also the Gold Award at the Hong Kong Global Design Awards, Hong Kong's Best Design etc.

临时性
Temporality

Design Academy Eindhoven graduate Tuomas Markunpoika Tolvanen has covered pieces of furniture in a fine web of steel rings before destroying them with fire. Calling his project Engineering Temporality, Markunpoika Tolvanen was inspired by his grandmother's disintegrating memories as she struggled with Alzheimer's disease.

Inspired from a personal agony within his family

Tuomas Markunpoika Tolvanen
"Engineering Temporality"
"工程临时性"

埃因霍芬设计学院毕业生托马斯·马昆波一卡·多尔瓦农(Tuomas Markunpoika Tolvanen)将一系列钢环覆盖在家具上面，然后将家具焚烧。马昆波一卡·多尔瓦农将他的项目叫做"工程临时性"是因为他的祖母受到阿尔兹海默氏病的困扰只有不完整的记忆。

从其家庭个人苦恼中获得灵感

家庭制造
Home manufacturing

Vadim Kibardin
"Black Paper 37"
"黑色纸张37"

这是一款采用37层纸板制作而成的扶手椅，将纸板一片一片地组合起来达到座椅所要求的高度。

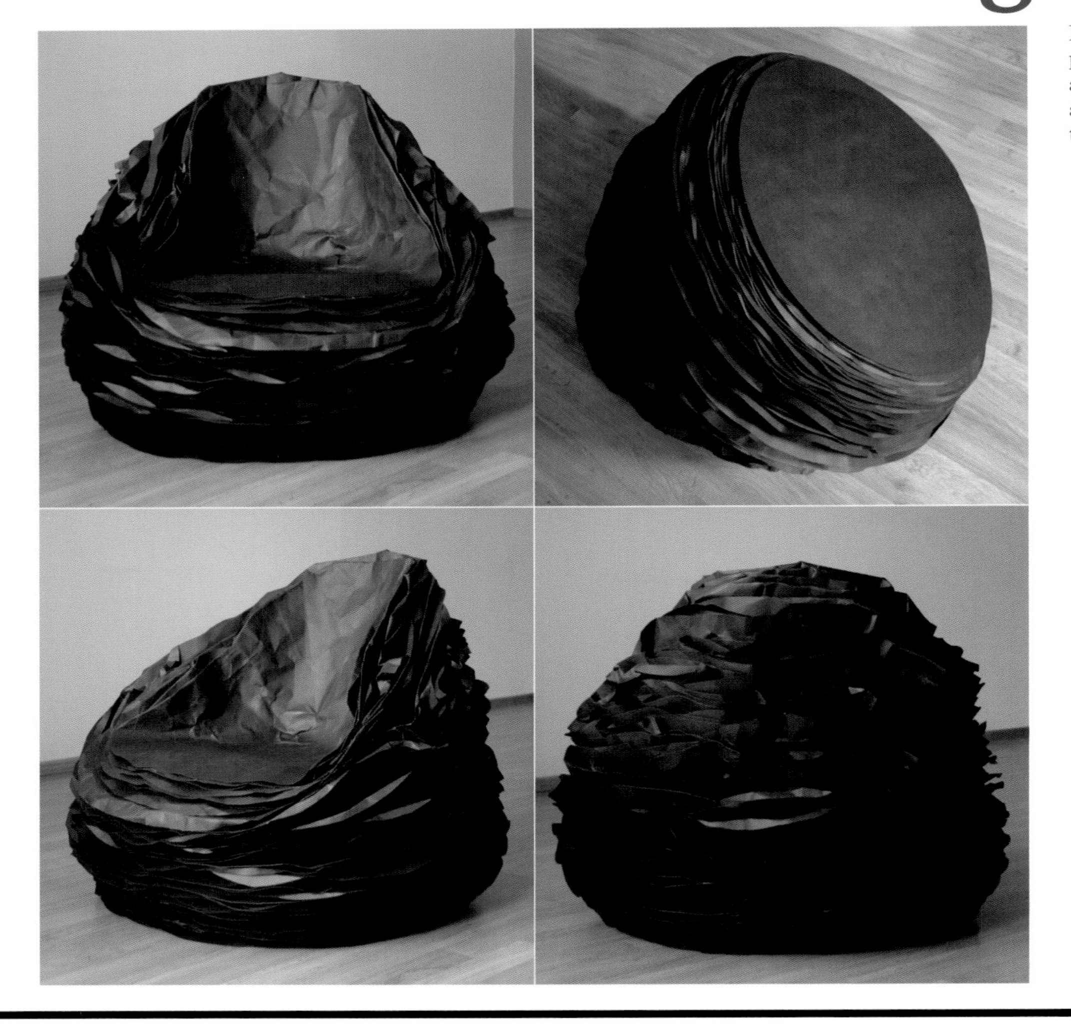

It is an armchair made of 37 paper layers and cardboard, arranging sheets one by one to achieve the required height of the chair.

"鲍莫斯2号"是一款创业沙发，模拟刚从树枝上掉落的树叶。手工弯曲的沙发钢架采用与叶脉相同的图案，充分展示出其自然结构和强度。

"叶脉"清晰可见

The Bomers' Nr. 2 is a sofa resembling a leaf just fallen from its tree. The hand curved steel frame of the sofa follows the same pattern as the veins of a leaf adapting its natural structure and strength.

The "veins" are visible

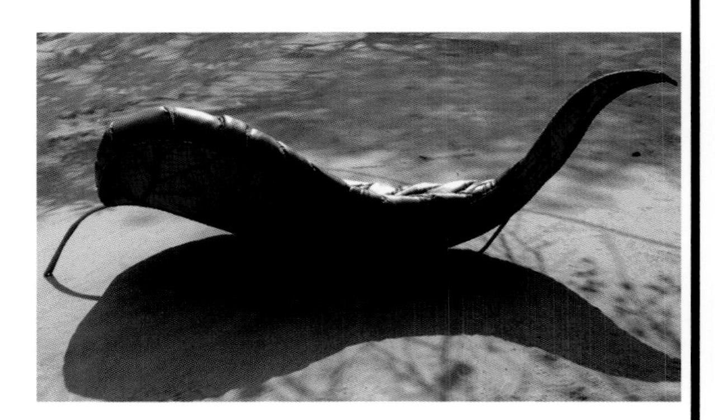

手工制作
Handcrafted

混凝土块拼接家具
Monolithic piece of furniture

Stefan Zwicky
"Concrete Chair"
"混凝土座椅"

瑞士设计师斯特凡·兹维奇(Stefano Zwicky)用钢筋和混凝土对勒·考尔布斯尔(Le Corbusier)设计的LC2扶手椅进行了重新创作。

Swiss designer Stefan Zwicky has re-created the iconic LC2 armchair by Le Corbusier with rebar and concrete.

一款软模具
A soft mould

Yael Tandler
"Baztek Stools"
"巴兹泰克凳子"

就像制作面包一样，这种制造技术已经成为一种奇妙而有趣的活动。

创造各种实验形状

这些物品采用家庭材料，如气球、橡胶板、木板或水泥等。

Like making bread, the manufacturing technique becomes a curious and playful act.

Creating experimental shapes

They use household materials: balloons, rubber bands, pieces of wood, cement...

混凝土强度
Strength of concrete

These benches are made by folding fabric that is impregnated with cement then drenching it in water.

It is simply a construction of canvas panels

The material combines the warm softness of the cloth and the stability of the cold concrete, but the finished surface keeps the soft appearance.

Florian Schmid
"Stitched Concrete Bench"
"缝制的混凝土长椅"

这些长椅由织品折叠而成，内部注入水泥后在水中浸泡成型。

这是一款帆布板制成的长椅

这款长椅将温暖柔软的布料特点与冷混凝土的稳定性完美结合在一起，而修饰后的表面依然保持柔软的外观。

大半径连续曲线
Sweeping curves

Foster + Partners
"Arc Table"
"弧形圆桌"

雕塑般的底座采用水泥和有机纤维制成，形成一个三条支腿的简洁结构。桌面是一块直径为140cm或150cm的圆形钢化玻璃盘。

The sculptural base, made from a material composed of cement and organic fibre, is a single form anchored by three legs. The top is a tempered glass disc, measuring either 140 or 150 centimetres in diameter.

简单的几何形状
Simple geometric shapes

Komplot Design
"Concrete Things"
"混凝土物品"

这是一系列户外混凝土家具，能够显示公共空间中个人与集体之间的关系。这些家具的变形表明曾经有人坐在这里。

A series of outdoor concrete furniture show the relation between the individual and the collective in public space. Its deformation keeps memory of somebody once seated in them.

实木家具
Solid wood

Eduardo Baroni
"Cordame Armchair"
"绳缆扶手椅"

该作品以老式木制玩具(悠悠球和剑玉)为设计依据，但其并不代表某一种特定的玩具设计，而是唤回设计师当时玩这些玩具时的感觉。

The design is based on old wood toys (yo-yos and kendamas), not to represent a specific toy design but to bring back the feelings that the designer had when handling these objects.

Continuum Fashion
"3D Printed Strvct Shoes"
"Strvct立体打印鞋"

美国连续时尚(Continuum Fashion)设计工作室推出了他们设计的第一个立体打印耐磨鞋系列。

American design studio Continuum Fashion have launched their first range of wearable 3D-printed shoe collection.

轻便舒适

这些打印鞋由三角网格面和漆皮内底衬里组成，鞋底涂有一层合成橡胶来增加鞋的附着摩擦力。

Extremely lightweight

They feature a triangulated lattice surface and lined with a patent leather inner sole, and coated with a synthetic rubber on the bottom to provide traction.

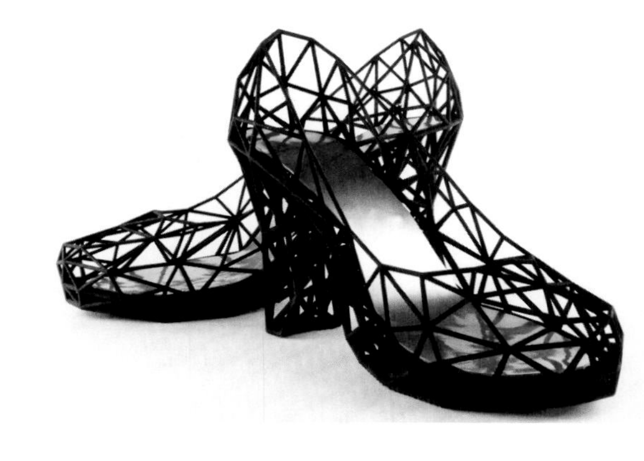

迪克 ·夏佩尔斯(Dik Scheepers)设计的“粗糙”(UnPolished)是一款采用轻型纸砖(一种由再生纸张纤维和水泥制成的建筑材料)和木材制成的家具。这是一种实验材料，具有成本低、用途广、质量轻、触感舒适等特点。

UnPølished, by Dik Scheepers is furniture made of lightweight papercrete, a construction material made of re-pulped paper fiber and cement, and wood. It is an experimental material with feature of low cost, versatile, light weighted, and comfortable.

实验作品
It is about experimentation

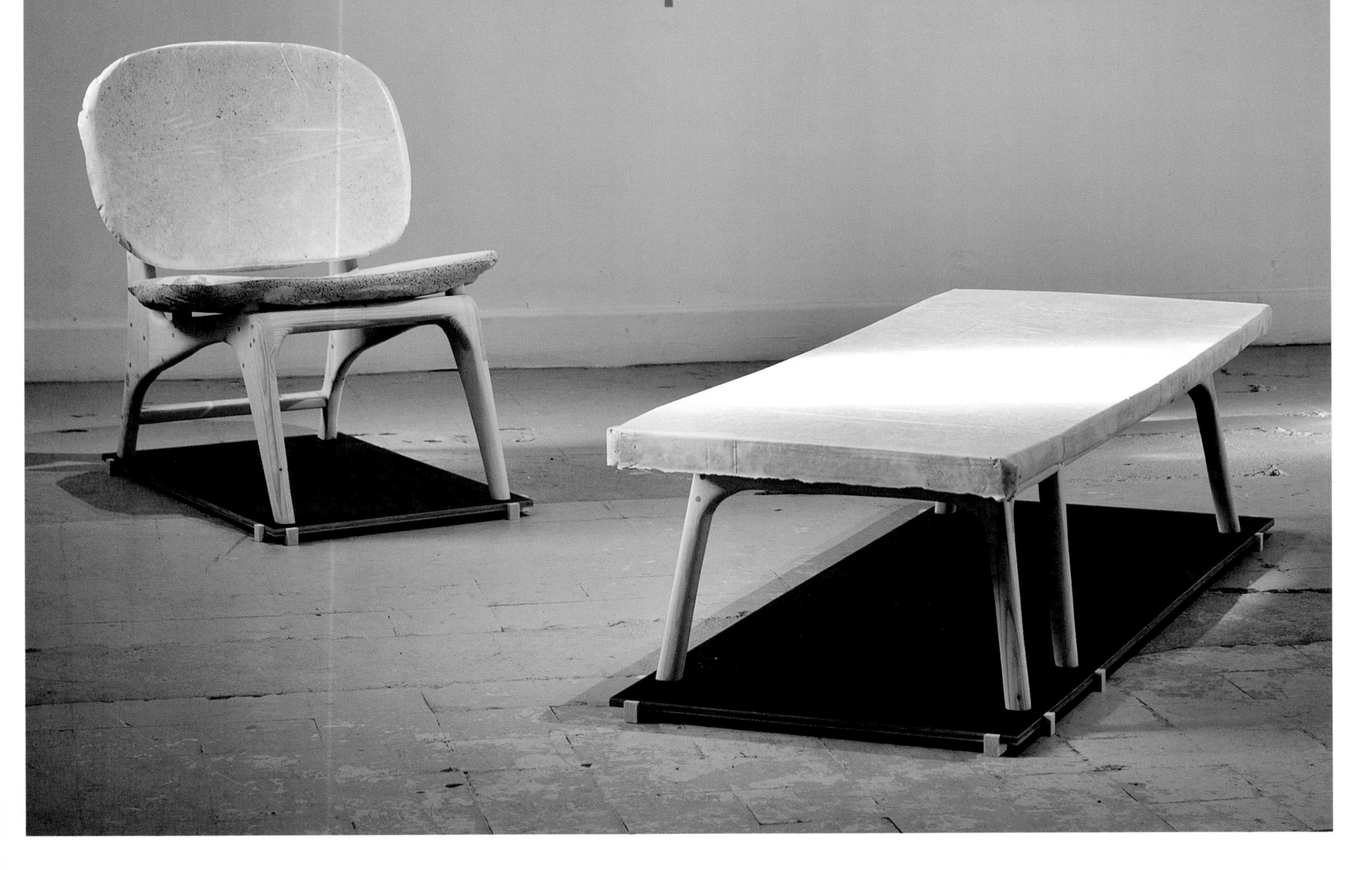

充满空气的空心实体
An empty solid but full of air

Lapo Germasi
"Inspiration chair"
"灵感座椅"

这是一款由PVC外套、低密度聚亚氨酯圆球和吸气阀门组成的座椅，具有多变的外观形状。

A cube of electrowelded PVC, balls of low-density polyurethane and a suction valve. A shape without shape.

Stefan Borselius
"Wilmer T "
"威尔默T"

多功能安乐椅

A multi-functional easy-chair

科尼利斯(Borselius)提高了我们坐在座椅上的舒适感。威尔默T将两块灰色实心面板以不同高度附在座椅的两侧，成功创造出一款几乎适用于所有场合的多功能座椅。

追求舒适度与灵活性

与其他具有规则表面的座椅相比，这款座椅占用的空间非常小，使用起来更加简单方便，适用于更多场合。

Borselius has improved how we spend our time seated in a chair. With two solid ash tabletops attached to its sides at different height levels, the Wilmer T produces a multi-functional platform ideal for potentially anything.

It seeks for comfort and flexibility

The chair extremely reduced space it occupies, compared to the one of a regular table surface, adds even more value to the simple convenience of performing multiple tasks.

奖项
AWARDS

© OMA / Dominik Gigler

© Iwan Baan

雷姆·库哈斯赢得詹克斯奖
Rem Koolhaas wins Jencks Award

雷姆·库哈斯(Rem Koolhaas)荣获“2012年度詹克斯奖”。“詹克斯奖”授予每年度为国际建筑理论和实践做出重大贡献的个人(或案例)，此次颁奖在位于伦敦的英国皇家建筑师学会举行。此次颁奖活动将由建筑理论学家查尔斯·詹克斯(Charles Jencks)主持，库哈斯将进行公开演讲。詹克斯评论说：“与其他同时代建筑师相比，雷姆·库哈斯在建筑理论与实践两方面均做出了重大贡献。”往届“詹克斯奖”获得者包括扎哈·哈迪德(Zaha Hadid)、FOA建筑师事务所、彼得·艾森曼(Peter Eisenman)、塞西尔·巴尔蒙德(Cecil Balmond)和斯蒂文·豪尔(Steven Holl)。

Rem Koolhaas has been named the winner of the Jencks Award for 2012. Given annually to an individual (or practice) that has recently made a major contribution internationally to both the theory and practice of architecture, the Jencks Award was presented at the Royal Institute of British Architects (RIBA) in London. The event will feature a public lecture by Koolhaas, chaired by Charles Jencks, architectural theorist. Jencks commented: “Rem Koolhaas, more than any other architect of his generation, has built a parallel life between the theory and practice of architecture.” Previous winners of the Jencks Award include Zaha Hadid, Foreign Office Architects, Peter Eisenman, Cecil Balmond and Steven Holl.

© Iwan Baan

汉斯 · 梅耶奖
Hannes Meyer Award

西班牙涅托 · 索伯加诺(Nieto Sobejano)建筑师事务所设计的莫里茨堡博物馆(萨勒河哈雷)项目被授予"2012年度汉斯 · 梅耶奖"。此次评审由包豪斯德绍基金会会长飞利浦 · 奥斯瓦特(Philipp Oswalt)主持，并且将此次奖项授予卡罗建筑师事务所设计的Salbke Lesezeichen露天图书馆项目。

此次参赛作品包括25家设计事务所过去五年内在萨克森 · 安哈尔特州(Sachsen Anhalt)完成的项目。

Spanish architects Nieto Sobejano have been awarded with Hannes Meyer Award 2012 for Moritzburg Museum (Halle, Saale). The jury chaired by Philipp Oswalt, director of the Bauhaus Dessau Foundation, awarded an ex-aequo prize to Karo Architekten for Salbke Lesezeichen.

There were 25 shortlisted offices which submitted projects completed in the state of Sachsen Anhalt during the past five years.

© 2008 Roland Halbe/Moritzburg Museum exterior view

© 2008 Roland Halbe/Moritzburg Museum exhibition hall

2012年度国际建筑奖
International Architecture Award 2012

波尔斯+威尔森(BOLLES+WILSON)建筑师事务所为德国艾尔维特的一家混凝土厂设计的新总部大楼项目荣获"2012年度国际建筑奖"。"国际建筑奖"每年由芝加哥雅典娜建筑与设计博物馆和欧洲建筑艺术设计与城市研究中心颁发，被认为是创新和优秀设计奖中最著名的国际奖项之一。

BOLLES + WILSON have received an International Architecture Award 2012 for their design of a new headquarter for a concrete plant in Erwitte, Germany. The International Architecture Award is annually awarded by the Chicago Athenaeum: Museum of Architecture and Design and the European Centre for Architecture Art Design and Urban Studies and is considered as one of the most prestigious international awards for innovative and outstanding design.

主编 · DIRECTORS AND PUBLISHERS
Gerardo Mingo Pinacho · Gerardo Mingo Martínez (西)

主办单位 · SPONSORS
西班牙未来建筑 · future arquitecturas s.l.

图形制作 · GRAPHIC PRODUCTION
Gerardo Mingo Martínez (西) · gerardo@arqfuture.com
赵磊 Zhao Lei · leizhao@arqfuture.com
左雯莎 Zuo Wensha · news@arqfuture.com
冯艳 Feng Yan · communication@arqfuture.com

中国地区公司合伙人 · CORPORATE PARTNER IN CHINA
赵磊 Zhao Lei · leizhao@arqfuture.com
地址 address: 中国杭州下城区文晖路303号
浙江交通集团大厦11楼
邮编 postal code: 310014
电话 telephone: +86 571 85303277
手机 cell phone: +86 13706505166

行政人员 · ADMINISTRATION
Belén Carballedo (西) · belen@arqfuture.com

FUTURE杂志社《万象建筑新闻》中国唯一指定合作伙伴
LEADING COLLABORATOR OF FUTURE ON PANORAMA IN CHINA
上海颂春文化传播有限公司
Shanghai Songchun Culture Communication Co., Ltd

销售部 · DISTRIBUTION DEPARTMENT
曾江福 Zeng Jiangfu
手机 cell phone: +86 13564489269
电话 telephone: +86 21 65877188

市场推广 · MARKETING DEPARTMENT
曾江河 Zeng Jianghe
手机 cell phone: +86 13564681595
电话 telephone: +86 21 65878760

翻译 · TRANSLATION
王坤 Wang Kun

广告 · ADVERTISING
china@arqfuture.com

future arquitecturas
Rafaela Bonilla 17, 28028
Madrid, Spain

www.arqfuture.com

未来建筑

future

ARQUITECTURAS